Feasts on the Farm

OVER 60 SEASONAL RECIPES AND STORIES OF SUSTAINABLE FARMING FROM TOMALES FARMSTEAD CREAMERY

Feasts on the Farm

By **TAMARA JO HICKS** & **JESSICA LYNN MACLEOD**

Foreword by **ALICE WATERS**

Photographs by **KATIE NEWBURN**
Food styling by **CHRISTINE WOLHEIM**
Illustrations by **JESSICA LYNN MACLEOD**

CHRONICLE BOOKS
SAN FRANCISCO

Text copyright © 2025 by Tamara Jo Hicks & Jessica Lynn MacLeod.
Photographs copyright © 2025 by Katie Newburn.
Illustrations copyright © 2025 by Jessica Lynn MacLeod.

All rights reserved. No part of this book may be reproduced in any form without written permission from the publisher.

Library of Congress Cataloging-in-Publication Data available.

ISBN 978-1-7972-2916-4

Manufactured in China.

Food styling by Christine Wolheim.
Design by Rachel Harrell.
Typesetting by Wynne Au-Yeung.

10 9 8 7 6 5 4 3 2 1

Chronicle books and gifts are available at special quantity discounts to corporations, professional associations, literacy programs, and other organizations. For details and discount information, please contact our premiums department at corporatesales@chroniclebooks.com or at 1-800-759-0190.

Chronicle Books LLC
680 Second Street
San Francisco, California 94107
www.chroniclebooks.com

Contents

Foreword 12

A Note from the Authors 14

Restoring Toluma Farms:
A History of the Land and
Dairying in Marin County 16

Forever Protected and Climate
Resiliency: Our Partners in
Sustainability 18

Walking the Talk 20

Next Gen: Teaching
Organic, Regenerative, and
Sustainable Practices 23

A Year on the Farm 26

Goats 28

Cheese 33

A Recipe for Renewal 38

SPRING

New Life: Kidding, Lambing,
and Green, Green Grass 45

To Start

Wild Mushroom, Thyme,
and Liwa Toasts 48

Peas, Cheese, and Caviar! 49

Asparagus and Teleeka Tart 51

Spicy Kale and Avocado Salad
with Farro and Koto'la 52

Endive, Pear, and Kenne Salad with
Toasted Walnuts and Honey 54

The Farm Healed Me 55

Eggs Brouillés with Truffle
Butter and Brioche 57

Childhood on the Farm 58

Mains

Farmer's Panzanella Salad 63

Savory Leek, Artichoke,
and Atika Tart 64

Liwa and Ham Frittata with
Garden Lettuces 66

Spring Vegetable Risotto with
Kenne, Atika, and Liwa 67

Roast Chicken with Fennel
and Olives 70

Marinated Lamb Chops 72

Slow-Braised Lamb Shanks
with Apricot Couscous 73

Next Gen 76

Sweet

Crisp Popovers with Goat Kefir and
Mulberry Rhubarb Compote 80

Meyer Lemon and Rosemary
Panna Cotta 81

Herbaceous Cocktail Cookies 83

Great-Grandma Opal's
Chocolate Sheet Cake 84

SUMMER

Summer Dinner Parties with Our
Favorite Chefs and Humans 89

To Start

The Perfect Summer
Cheese Board 93

Roasted Sungold Tomato
and Teleeka Bites 94

Whipped Garden Herb and
Liwa Mousse 96

Kale Salad with Cashews
and Koto'la 97

Charred Watermelon Salad
with Basil and Koto'la 99

Stone Fruit and Flowers 100

Smoked Trout on Crackers
with Liwa and Olives 102

The Cheesemaker's Proja 103

Mains

Toluma Ploughman's Lunch 108

Dijon Grilled Cheese 109

Heirloom Tomato and
Nasturtium Galette 110

Balsamic Salad Pizza 113

Meyer Lemon and
Honey Pizza 115

Peach and Serrano
Ham Pizza 116

Lemon Bucatini with Atika 119

Teleeka and Summer
Squash Lasagna 120

Truffle Roast Chicken 125

Cornish Game Hen Dijonnaise 126

An Alternative View of Cheese: Cultural Context Matters 127

Stemple Creek Rib-Eye with Rosemary Navy Beans 131

Sweet

Tahitian Vanilla–Cardamom Sheep Milk Ice Cream 134

Mid-August Blackberry Corn Muffins 137

Californios Banana Cachapa 138

The Grace of Goats 139

Liwa Basque Cheesecake with Honeyed Nectarines and Nasturtiums 143

Rhubarb Crumble Cake 144

AUTUMN

Friendsgiving: Our Community of Women-Run Farms 148

To Start

Warm Olives with Atika, Herbs, and Lemon Zest 153

Goat Cheese Tatin 154

Farmers Are Conservationists 159

Roasted Harissa Root Vegetables with Koto'la and Toasted Pumpkin Seeds 160

Goat Cheese Croque Monsieur Canapé 162

Toasted Liwa Muffin with Dill and Honey 163

Potato Leek Soup with Smoked Pepper and Liwa Crema 164

No, Goats Don't Eat Tin Cans 165

Mains

Maple-Crusted Butternut Squash with Liwa 169

Smoky Baked Eggs with Chickpeas and Goat Cheese 170

Ross's Niçoise Salad 173

Harvest Lentil Soup with Fried Liwa and Butter 174

Sweet

Gravenstein Apple and Marmalade Grilled Cheese 179

Bay Laurel and Fig Clafoutis 180

WINTER

Season of Slow Living 185

To Start

Smoked Trout on Toast with Walnut-Lemon Gremolata 189

Roasted Beet Salad with Teleeka and Pancetta 190

Little Gem and Watermelon Radish Salad with Koto'la 192

Farm-Stay Fun 193

Mains

Wild Mushroom and Teleeka Quiche 196

Whipped Liwa BLT 197

Roasted Dungeness Crab with Citrus Herb Butter 199

Fennel, Avocado, and Chicory Salad with Citrus and Atika 200

Doug's Balsamic Fig and Tomato Chicken 203

Kimchi, Kenne, and Bacon Pajeon 204

Lamb and Pork Sausage Cassoulet 205

Sweet

Campari Hazelnut Cake 208

Tomales Farmstead Creamery Soufflé 211

Giving Thanks: The Team at Toluma Farms and Tomales Farmstead Creamery 214

Acknowledgments 217

Index 220

Foreword

One of the central tenets of Edible Education, the curriculum used in all the Edible Schoolyard Projects around the world, is that "beauty is a language of care." When a classroom (or any place, for that matter) is beautiful and orderly, students feel, from the moment they enter, that they are cared for, that someone has taken the time to consider and create a space just for them, and that they are honored and supported, which inspires them to learn and grow humanely.

The converse can also be true: that caring creates beauty. If one keeps their books in order, for example, that shelf is beautiful. If one makes their bed well, that room is welcoming. If one carefully picks a flower to put on a teacher's desk— even though it's a weed!—it's beautiful. It's the intention that makes it that way. Caring implies a certain attention to what's outside yourself and how you can nurture it. Caring implies stewardship.

More than anything, Toluma Farms is a beautiful place. And I don't mean just environmentally, which it is, but in all aspects. The way the fields are tended regeneratively; the integration of the animals into farm life; the attention and responsiveness to weather patterns, growth cycles, and the lay of the land; the evolving craftsmanship of their cheesemaking; the connections to their community; their respect for workers and visitors; the deliciousness and integrity of their cooking; the commitment to education, both their own and for others—all of this is what makes up true stewardship of the highest and most respectful order. The pages of this book have captured the beauty in that stewardship—that universal language of care.

Alice Waters

A Note from the Authors

"Eating is an agricultural act," says writer, environmentalist, and farmer Wendell Berry. Here on Toluma Farms in Tomales, California, we take those words to heart. The farm is a place where the earth tells tales of resilience and each season unfolds its unique gifts. As coauthors of this cookbook, our journey is a blending of passions and experiences, weaving together the threads of sustainable farming, culinary exploration, and a determination to be radically optimistic about the future of our planet and future generations.

I, Tamara, share my family's story as the stewards of this unceded Coast Miwok land, where restoring ecosystems is a way of life. Through essays from my family and members of the community on regenerative agriculture, cheese making, and our successful apprenticeship program, we share the story of our farm—a haven where the next generation of regenerative farmers is nurtured and empowered to care for the future of our planet and one another.

And I, Jessica, bring to you simple recipes that use seasonal ingredients and a reverence for nostalgic cooking. Having spent time as a volunteer on Toluma Farms, working closely with the herd managers and immersed in the barn's slow and peaceful energy, my food philosophy is deeply rooted in letting the ingredients shine when they're most abundant, preserving the techniques and flavors of past generations, and celebrating the joys of bringing people together around a comforting and celebratory meal.

Together, we invite you to embark on a seasonal culinary journey inspired by the heart and history of the farm. From the pastures teeming with life to the farmhouse kitchen alive with aromas, these recipes and stories are a testament to the abundance of our land and the bonds of community that sustain us. This cookbook is a tribute to the countless shared meals, the vision of local, sustainable agriculture, and the unwavering support of our community.

Welcome to our table,

Tamara and Jessica

Restoring Toluma Farms: A History of the Land and Dairying in Marin County

In the winter of 2003, our family purchased a 160-acre parcel of land in West Marin County, California. Nestled in the small hamlet of Tomales (pop. 203), just 60 miles north of San Francisco, it had been a cow dairy for more than eighty years.

As with many dairies throughout the country in the eighties and nineties, it had gone out of business due to the rise of big dairy, and the existing structures had fallen into disrepair. Our family, including my husband, David, and our daughters, Emmy, Josy, and Megan, took on the challenge of bringing it back to life, and for over twenty years now we have stewarded this magical Pacific Coast land that we call Toluma Farms.

Marin County's rich history of dairies began around the time of the Gold Rush. It took the foresight, creativity, and intelligence of a woman named Clara Steele to inspire an endeavor that would change the course of history in California. Clara Steele had settled in Sonoma County in the mid-1850s with her husband and family. Like hundreds of others, the Steeles had tried their hand at mining, but they were ultimately drawn to the verdant Bay Area hills. On a visit to the vibrant new city of San Francisco, Clara quickly noticed the need for fresh dairy products. In 1857, she hired a local Native American man (name unknown)—most likely Coast Miwok, a tribe indigenous to Marin—to rope some wild cattle that grazed near her home. She milked these cows and, using her English grandmother's recipe, began producing cheddar cheese. Clara was not the first Marin settler to milk a cow or make butter and cheese, but she was the first in Marin to transform dairying into a business. In San Francisco, she found eager buyers.

That same year, the Steeles signed a lease on land in Point Reyes and invested in the creation of a 10,000-acre dairy ranch. Here, pastures bathed in fog created conditions they termed "cow heaven." Within a few years, the Steele family was operating three busy dairy ranches and a schooner that made regular deliveries to San Francisco. They had successfully built the first large-scale dairy business in California, and others quickly followed.

By 1862, Marin County was producing a quarter of California's butter. Immigrants predominantly from Ireland, the Italian-Swiss mountain country, Portugal, and the Azores established family farms that remain vital to our local economy today. Towns like Tomales, Olema, and Nicasio became early trading centers for the county's growing dairy ranches.

FEASTS ON THE FARM

At their peak, there were more than three hundred dairies in Marin County. Today there are fewer than thirty. Each year, more and more dairies shut down due to obstacles spanning not only the rise of big dairy but also a yearslong drought, rising fuel and feed prices, a lack of succession plans, and tougher regulations. The future of the dairy industry in Marin and Sonoma counties is uncertain, but we hope to inspire future generations to keep the essential practice of sustainable farming going for years to come. This book is a physical manifestation of our desire to record and share our farming practices and philosophies for the sake of posterity.

Forever Protected and Climate Resiliency: Our Partners in Sustainability

One of our best decisions was simply starting up a farm, in Marin County in particular. But at the outset, we didn't fully know the richness of resources for the farming and food community. Without the support of those partners, our gates would have been closed many years ago.

A week after we closed on the property in 2003, we called Natural Resource and Conservation Services (NRCS), as our realtor had said only, "Good luck, and you should call this number." (I think all realtors should offer this resource to all agricultural property buyers.) NRCS is an arm of the USDA that provides financial and technical resources for farms. Because of NRCS, we have improved our soil by diverting runoff, restoring riparian areas, removing those ten thousand tires, planting trees and bushes for windbreaks, and so much more. We have been working with many of the same trusted people for over fifteen years, and they know the land as well as we do.

We also began working with our local Resource and Conservation District (Marin RCD) and Point Blue Conservation, which also runs Students and Teachers Restoring A Watershed (STRAW). We have always felt supported by these organizations, never judged for our lack of knowledge about bird or tree species or how to interpret the data for a soil sample.

The need is great, and urgent. Across the country as a whole, we lose a staggering 175 acres of farmland every hour. So from the beginning of our starting the farm, we knew we wanted the land in a trust to ensure that it looks like it does now in perpetuity. We chose Marin Agriculture Land Trust (MALT), as it was important to us that agriculture remain active on the land forever, continuing to be a robust part of the local economy. MALT was founded in 1980 by two pioneering women conservationists, Ellen Straus and Phyllis Faber, working in partnership with the ranching and farming community. Of the total 110,000 farmable acres in Marin, 60,000 acres are now protected in the trust (that's twice the area of the city of San Francisco!), representing close to one hundred family farms. We wanted to play a small part in this effort.

As meaningful as it is knowing that this land will remain farmland into the future, we also realize that land trust and conservation organizations historically excluded people of color, women, and the LGBTQ+ communities, either through policies that excluded marginalized groups or by viewing farmland

protection and conservation as separate from issues of equity and social justice. Our farm will continue doing the work to ensure that we include all the voices in our community, and we'll partner with the organizations that share similar values and missions.

The incredible individuals who work at our partnership organizations have taught us so much about climate-resilient farming. After twenty years, it is automatic for us to run every decision through the filter of how it will affect the land, people, animals, and community. Thanks to the Carbon Cycle Institute that assisted us with a Carbon Farm Plan ten years ago, we have a road map to guide us in best practices for our farm.

In our years on the farm, we have also benefited from learning about our natural environment as it has been developed and stewarded by those who have lived here far longer than we have. In reading M. Kat Anderson's wonderful *Tending the Wild: Native American Knowledge and the Management of California's Natural Resources*, I was really struck by my misconception about what is "wild" in our state. Naturalists such as John Muir were early proponents of a faulty view that persists today—that much of California was pristine, untouched wilderness before the arrival of Europeans. Anderson underscores that what "Muir was seeing when he admired the vistas of Yosemite and the Central Valley were actually the fertile gardens of the Sierra Miwok and the Valley Indians, modified and made productive by centuries of harvesting, tilling, sowing, pruning and burning."

The oversimplified stereotype of the hunter-gatherer misses the complex picture that has emerged from historical material: California's indigenous people were and are active agents of environmental change and stewardship. This ecological knowledge is essential if we are to find our way back to a more sustainable way of managing working and nonworking lands alike.

In 2023, Greg Sarris—a local tribal member and chairperson of the Federated Indians of Graton Rancheria—and Obi Kaufmann—a naturalist, writer, and illustrator—started a podcast called *Place and Purpose*. The storytellers state that the chaos of the modern world is tempered by a deep connection to land, home, and community. Their informal discussions on the topics of land, environment, community, family, and more inspire me to continue to look outside of our 160 acres for the support and knowledge that awaits.

Walking the Talk

At Toluma Farms, we engage in a number of sustainability practices to make sure we are part of the solution and not the problem. Though milking twice a day is standard practice at conventional modern dairies, we milk just once a day. This not only reduces our use of water and electricity but also ensures that our team can be home in the evenings or go out and have a life outside work. We don't ship our cheese out of California, as we believe one of the best things you can do for the planet is reduce your carbon footprint by eating locally. There is so much regionally wonderful artisan cheese to be consumed, it seems silly to have a Vermonter eating our cheese in Vermont—though we're happy to have them eat it when they visit.

After we had our land certified organic, we began a multi-year planting program. We aim to plant one hundred trees or bushes a year, and thus far we have met that goal. In the last twenty years, we have planted over one thousand trees and bushes. We plant for many reasons; our coastal plains area was so appealing for dairying a century ago in part because there were few naturally established trees—unlike nearby coast range areas that are densely forested with redwoods and other conifers, like Occidental to our north. So our area's good for grazing, but also insanely windy for much of the year. The newly arrived dairy farmers soon realized they needed windbreaks! Trees and bushes also help slow down rainwater runoff, allowing it to penetrate the soil rather than remaining on the top, giving us much healthier soil to support the pasture grasses that nourish our goats and sheep. We also plant to increase the biodiversity on our farm, providing habitat for birds, bugs, butterflies, and mammals.

The goats allow us to practice prescribed grazing. Livestock are naturally *selective* about the plants they eat; they tend to repeatedly graze some plants and ignore others. This is one reason we have multiple species (goats, sheep, and chickens) in the pastures. Selective grazing weakens the more desirable plants and allows unwanted plants to thrive and multiply. Nearly all pastures have areas where livestock concentrate, such as around water, bedding grounds, and feed grounds. If the pasture is used continuously, these areas become overused, and the pasture quality deteriorates. Thus, you will see animal husbandry and land managers Audrey Raine and Julia Brandt planning the winter and spring daily pasture moves, and in the summer and fall, every-other-day moves. Audrey, Julia, and the team utilize electric fencing, tracking collars, and a herding dog named Molly to successfully move the animals around.

Audrey, Julia, and the team's intensive planning gives more desirable plants a chance to grow and multiply. According to NRCS, prescribed grazing increases the number of high-quality plants per acre, "improves wildlife habitat, reduces soil erosion, and conserves water. By resting pastures, over-used areas are allowed to become productive."

We installed a 15,000-gallon water catchment system off of our barn to capture rainwater that we can use to irrigate pastures, thereby extending the grazing season, and to have on hand for fire mitigation. Our next big sustainability project will be installing solar to decrease our use of fossil fuels and lower our utility bills.

Next Gen: Teaching Organic, Regenerative, and Sustainable Practices

None of our communities can continue to thrive without a passionate and dedicated next generation of farmers, ranchers, and food makers, yet our country still has a long way to go to ensure that these future generations have a fighting chance at keeping the farms alive. We need creative financing options for purchasing land, an overhaul of the current permitting process to expedite the construction of critical farmworker housing, and innovative collaborations across sectors. Without our immediate care, attention, and action, we'll continue to see the decline of small family farms throughout our country and the world.

In David's and my respective fields of medicine and psychology, teaching the next generation is an integral and respected part of the work. Our mentors and teachers quickly turn into colleagues and friends who champion us along the way. They serve as a resource for solving complex problems, and they send patients our way because of the deep trust developed over the years.

We knew we wanted to mimic this model of training and development at Toluma Farms, so from the very beginning, we prioritized a robust apprenticeship program. Our apprentices live at the farm for six months. During their tenure, we provide comprehensive hands-on training and education in animal husbandry, soil health, climate-resilient farming, grant writing, agritourism, bookkeeping, social media, and many other aspects of managing a small-scale farm and cheesemaking operation. Each apprentice spends four days a week on the farm and one day in the creamery making cheese. We are fortunate to have had the opportunity to learn from all the talented, inquisitive apprentices who have brought innovative ideas and experiences to our land.

As David and I began to educate ourselves ahead of this wild adventure, we quickly realized there are very few small-scale, sustainable dairies making cheese in this country. Many dairies are either hobby farms or large-scale operations that we didn't view as environmentally sustainable. Even less common are farms operated by young people, women, BIPOC, or LGBTQ+ communities. We knew early on that we wanted to offer opportunities for marginalized populations to come to Toluma Farms and learn alongside us.

We've now hosted nearly fifty apprentices, many of whom we maintain close friendships with; they have enriched our farm and lives immensely. Today, our apprentices are spread across the United States and practice regenerative farming in Alaska,

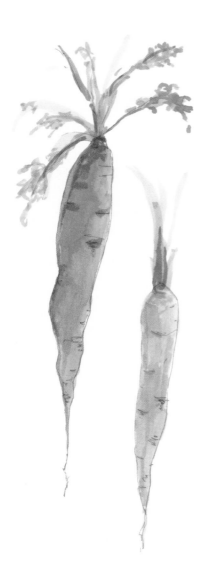

Arizona, Louisiana, Maine, Vermont, Illinois, Wisconsin, Washington, and New York. It brings us tremendous joy to see the work that these hardworking, creative individuals are pioneering.

Our robust apprenticeship program amply demonstrates that the next generation is ready to get their hands dirty. We'd like you to meet just a few of them.

Apprentice Sarah Roland came to Toluma Farms thinking she wanted to add goats to her family's third-generation property and start a dairy-cheesemaking operation. On the last day of her apprenticeship, we visited Double 8, the buffalo dairy of our friend Andrew Zlot, and Sarah instantly fell in love with these majestic animals. Within a few years, she opened Bayou Sarah Farms in Louisiana and now has a dozen water buffalo that are grass fed 365 days a year.

We've made a few trips to Sarah's farm and cheesemaking operation just outside of Baton Rouge. The Louisiana rain suits these buffalo well, and they spend much of their days in ponds around Sarah's property. Her farm is highly diversified: She not only makes buffalo mozzarella—one of the few operations in the country making this cheese—but also raises chickens, grows blueberries, and harvests honey. Her farm is progressive for Louisiana, where Roundup and mono crops are still the norm. In an amazingly short period, Sarah has become a mentor and model of how to farm regeneratively.

Another team member, Emily Eckhardt, worked creatively with Maine Farmland Trust to purchase a farm that had been in one family for generations but lacked a next generation of owners to take it over. Emily built a creamery on her property, wisely deciding to purchase milk from a neighbor rather than starting the complex operation of maintaining a herd of animals. Thus began Emily's venture into making elevated ice cream sandwiches under the name Afterglow Ice Cream. She supports her neighboring farmers by sourcing the milk locally and using locally milled flour for the sandwich cookies. Emily sells these amazing ice cream bars at farmers' markets in and near Embden, Maine, as well as in several small shops in her county.

A few of our apprentices who grew up in the Bay Area and attended University of California schools have become permanent parts of our team in West Marin. Jenna Coughlin grew up in Marin County and is a member of the Coast Miwok

Tribe, the original inhabitants on this unceded land. Jenna decided at the end of her apprenticeship that she wanted to challenge herself to become a cheesemaker. In no time, she was managing the creamery and making phenomenal cheeses. Jenna moved on to beginning her own business (Shepherds of the Coast), which utilizes goats and sheep for small-scale prescribed grazing to decrease the risk of fire.

Another Bay Area native, Nick Campbell, apprenticed at the farm and took a position making cheese in our creamery. He is now our head cheesemaker and lives in Tomales. Nick's mother grew up in Montenegro, and Nick brings to the farm his grandmother's and mother's food traditions through his weekly Friday Farmily meals.

If we all want to keep eating that incredible farmstead buffalo mozzarella, that fantastic ice cream sandwich, the soft-ripened triple milk cheese—all made with regenerative farming practices mere miles from where we can purchase these delights—we must find meaningful ways to support the next generation of farmers, ranchers, and food makers. You don't have to own a farm to be a part of the solution— you can simply visit one! We host hundreds of people on our farm every year for tours and events, and many of the neighboring farms do the same. We also partner with a tour company (Food & Farm Tours) that shares the same values of educating folks about regenerative farming. We are grateful to have the owner, Alex Fox, as a champion of our practices and delicious cheese! You can support local, state, and federal legislation and use your vote to support regenerative farming; the Farm Bill, renewed every five years, is probably one of the most significant funding sources for small-scale farming. You can get to know your local farmers by shopping at your community farmers' market, where 90 percent of your dollar goes directly to the farmer, compared with 30 to 50 percent wholesale. It is crucial to show up at these farmers' markets even when it is raining, cold, or windy, as the farmer is there, and the food has already been harvested or prepared to sell. Or, if you want to really get to know your community of farmers, you can volunteer at a local farm, as the project list never ends.

A Year on the Farm

Occasionally, you hear people say there are no seasons in the Bay Area. It *is* true that the changes are not as dramatic as those in the Midwest or East Coast, but there are distinct and beautiful seasons here. From late November to June, the farm's grass is as green as Ireland's. Starting around July through October and up to the first solid rains, the hills are a tawny gold. Some Bay Area visitors from afar think those golden hills are caused by the drought, but they have always been golden during the summer and fall, as celebrated in the Kate Wolf song "The Redtail Hawk":

> *The redtail hawk writes songs across the sky*
> *There's music in the waters flowing by*
> *And you can hear a song each time the wind sighs*
> *In the golden rolling hills of California*

In late January, the first thing to bloom on and around the farm are the pussy willow trees, which are tall and slender with fuzzy silver buds. We have many willow trees in Marin County as they are the ideal tree to plant in riparian areas to heal the land. Willow trees are planted along creek beds and used for habitat restoration and erosion control. They grow quickly and establish strong root systems that filter toxins. We mostly plant the arroyo and Sitka willow species, as they are native to Northern California. Willows root easily when you plant a fresh cutting, usually in early December to early March.

Our first restoration project of planting willows was in a ravine filled with old tires—ten thousand of them, as it turned out! We pulled out all those tires and worked on the plantings with STRAW. More than fifteen years later, the riparian area is thriving, and our goats love to occasionally flash graze the trees. We have also planted willows that the Coast Miwok tribe can use for their basket making; tribal members have taught us how to prune and maintain the trees for this purpose.

In mid-February, we are surrounded by vibrant yellow from wild mustard. All the parts of the wild mustard plant can be eaten at any point in its development. This is one of the first blooms that ends up on our winter cheese boards. Just like humans, no two cheese boards are exactly alike, as our cheeses change seasonally, and the edible plants and flowers we add to our boards from the farm also change from day to day.

By March, nature has gifted us with hillsides that look like a painting. Of course, we have our state flower, the orange California poppy (a common old wives' tale that gets passed down from kid to kid is that they will be arrested if they pick one!). The hills and fields are also carpeted with other flowers, including clovers of every color, larkspur, buttercups, lupine, tidy tips, and nasturtiums.

Summer brings thistle, an unwanted plant for cow and sheep ranchers, but amazingly goats can and will eat the spiky plant. It has nutritional value for them as well, so it is a big win all around. We then are graced with Queen Anne's lace and, unfortunately, hemlock, which is toxic to goats (as well as humans) if eaten.

In the fall, we have wild blackberries everywhere, another favorite of our goats and human children, and our small orchard of heritage apples and pears for our cheese boards, snacks, cider, and pies.

The seasonal signs include more than just plants and flowers, of course. Barn swallows arrive in May, delighting us with their swift swooping flight, and leave in August after raising their broods in mud nests along the eaves of the barn. On long summer days, we revel in the glory of the Pacific Coast, just a few miles from the farm, where we can easily go for lunch or a walk at the end of the day. December brings the start of the massive southern migration of around twenty thousand California gray whales from the Arctic down along the Pacific coast toward Mexico. Dungeness crab season is typically November through January.

These are the natural wonders. But of course, as farmers, we plant, grow, and harvest—and we are fortunate in the Bay Area to have something planted in the ground year-round.

So yes, without looking at a calendar, we clearly know what month it is and look forward to the bounty each season brings. And as you cook Jessica's delicious dishes, you will get to know, season by season, what is fresh and ripe in our garden, our neighboring farms, our friends' gardens, and, of course, the farmers' markets.

Goats

Goats to the Rescue: Vegetation Management and Wildfires

Many of our visitors who come for a tour, cheese tasting, or farm stay ask us why we chose goats, especially since cows dominate the area's farms, and ours was formerly a cow dairy. By the end of their tour or stay, they are no longer asking; in fact, they are trying to figure out how they can get two goats in their car to take home. Goats really are the G.O.A.T (greatest of all time)—they're highly intelligent and very social animals. As a psychologist, I believe they are the ultimate emotional support animal. The kids fall asleep in your arms and will be content for hours. The does and bucks will follow you around, wanting to know what you're doing and to be by your side. When they lose interest, they meander off on their own and can be very happy in their herd—but they hate being by themselves. So we chose goats mostly because they bring us and the hundreds of people who visit each year endless joy. And let's be honest: We could all use more joy in our lives. Finally, David and I are both Capricorns, so we feel a sort of goat kinship!

But we also chose goats for more than just love. Right before we started our farm twenty years ago, we had read Al Gore's *An Inconvenient Truth* and Bill McKibben's *The End of Nature*. We were young at the time, and David and I became concerned as to what climate change crisis our young girls, Josy and Emmy, would be inheriting. We didn't want a farm that contributed to the problem. We wanted a farm that was part of the solution—and sustainably, goats made the most sense to us.

Goats are browsers, not grazers (more on the difference in a moment). They like woody things, which is why they are used for private and public lands for grazing. They will eat the poison oak, thistle, and blackberries that cows and sheep won't eat. Our farm has several waterside hedgerows and fields of native and wild shrubs and trees that provide habitat for birds, butterflies, worms, bugs, moles, badgers, deer, coyotes, and even an occasional mountain lion. We do not use these areas as permanent pastures but will let the goats roam there a few times a year for a couple of days so they can manage the vegetation.

Each year, once our male goats are weaned, they are rehomed to various weed management businesses or other ranches and vineyards that want to control invasive plants without using weed whackers (which are not only noisy and

polluting but can spark fires). This practice goes back centuries, but in the modern agriculture era, it has been pushed aside by machines and chemical herbicides. As West Coast wildfires have become a year-round threat in the twenty-first century—ravaging over 400,000 acres of land annually in recent years—goats have become part of California's strategy to reduce wildfire risk. It's incredibly economical to use goats instead of a brush crew of people using fuel-dependent power tools and heavy equipment. The goats need only water, mineral blocks, and a livestock guardian dog to ward off coyotes and mountain lions.

Happy and Healthy Goats and Sheep for Quality Milk and Fiber

There are some real misconceptions about goats, so let's get them out of the way. Those of us who grew up watching *Looney Tunes* learned that goats eat everything, including tin cans. Now that we have a farm with over a hundred goats, I have learned that goats intuitively know what they need, but they will explore edible and non-edible items out of curiosity. Unfortunately, many people have experienced goats being aggressive in small enclosures, like petting zoos, where they are not fed a balanced diet. But when goats are treated with kindness and given space to roam and explore, they are extremely friendly, loving, and gentle creatures—very similar to dogs in the way they interact with and adore people. These happy, healthy goats produce high-quality milk, which in turn produces award-winning cheeses.

Using our goats' milk for cheese connects us with a rich history. Humans' use of goat cheese dates back thousands of years. It is believed to have originated in the Mediterranean region and can be traced back to ancient civilizations such as the Egyptians and Greeks. It has always been a peasant food, consumed by all.

Sadly, many people have had only goat cheese made with milk that was not fresh (i.e., wasn't sourced from the same property nor a nearby community). This often leaves people believing all goat cheese tastes so strong it's like you are licking a buck (a male goat). Often when people visit, we will give them two glasses of milk—one goat and one cow—for a blind tasting, and they really struggle to identify which is which. Fresh goat milk and fresh cheese should taste fresh and not like a barnyard.

Goat dairy foods not only taste great but are more digestible than cow dairy. The protein (called casein) in goat milk has a different composition than the casein in cow milk; it's more similar to human milk. Goat milk's protein complexes are smaller, meaning our digestive enzymes can make shorter work of them. No wonder goat cheese is the most commonly consumed cheese in the world.

Our farm is more focused on goats than on sheep as we began with goats, and their personalities and physical appearance make them a bit more distinctive than the sheep. Having said that, we do love our sheep. In the last couple of years, we have been moving toward having fiber sheep, such as Finnsheep, rather than the East Friesian dairy sheep that we began with in 2014.

We currently have a flock of Finn and Romney sheep and are excited to be working more closely with our vibrant local fibershed community in Marin, as exemplified by the Fibershed Learning Center on Black Mountain in Point Reyes and West Marin Wool Shed in Point Reyes Station. What's a fibershed? One blogger defines it as "a geographical landscape that provides the resources and infrastructure to create local fabric (local fiber, local dyes, local labor)." Through the years, many of us on the farm have taken their wool-spinning, embroidery, and dyeing classes. Their mission is to develop regional fiber systems that build soil and protect the health of our biosphere. Visitors are welcome to check out the Fibershed Learning Center, meander through their dye garden, and choose from their array of classes.

Our current wool projects are making organic dog beds and using wool for mulching around our trees and shrubs. Laying down raw wool in our garden and around our windbreak trees and heritage apple orchard retains moisture and stops weeds from sprouting—two goals of mulching. Straw mulch breaks down too quickly, leaving soil exposed and allowing weeds to grow, whereas "woolch," as we call it, lasts for a minimum of two years. This 100-percent natural material biodegrades back into the earth once it has served its mulching purpose.

Those in the fiber arts in Marin and Sonoma County are fortunate to be able to sell their products at the Wool Shed, which offers unique woolly goods that are California grown and climate beneficial.

Cheese

**Welcome to the Creamery.
Blessed Are the Cheesemakers!**

Ten years after we began Toluma Farms, we expanded into a new enterprise, Tomales Farmstead Creamery. Since 2013, we have been making cheese on the farm, and we have been blessed with extraordinary cheesemakers.

Making cheese on the farm depends on the seasons and our animals' natural reproductive cycles. No two cheese makes are exactly alike, so we can't just follow a recipe. An artisan cheesemaker needs to be part scientist and part artist.

David and I knew what types of cheeses we liked to eat, but in the beginning, we didn't know the first thing about making them. A weeklong cheesemaking class from Three Shepherds Farm in Vermont was enlightening and made us realize we needed to hire an experienced cheesemaker. We were extraordinarily lucky to find Anne Marie Vandendriessche, who had spent time in France learning cheesemaking. Together we figured out what we wanted to make: a fresh chèvre; a soft-ripened, goat milk, geotrichum-rind cheese; a mixed-milk (goat, sheep, and cow) soft-ripened cheese based on La Tur, a cheese we fell in love with in Italy; and an aged Spanish-style cheese.

Years later, cheesemaker Jenny MacKenzie joined us (coming from our friends' Redwood Hill Creamery) and added our goat feta-style cheese (the Greeks have proprietary ownership of the name *feta*). A third grader who visited the farm once said, "It is a fetish cheese!" When Jenny moved to Maine, Jenna Coughlin took over as head cheesemaker and added Bossy (named for badass bossy women), a soft-ripened cow's milk cheese, and Greco, a feta-style cow's milk cheese.

Our land is on unceded Coast Miwok land, so most of our cheeses have Coast Miwok names. We use the names to remind the thousands of people who visit us and eat our cheeses each year whose land we are on. We are grateful that the tribe is an active one in Marin and Sonoma Counties and we can engage with them on riparian projects and the rematriation of found objects such as a mortar bowl. Jenna is Coast Miwok and a wonderful liaison between us and the tribe.

We are tremendously proud of our delicious, award-winning cheeses! Three factors make them so wonderful: our focus on soil health, our outstanding animal husbandry, and our farmstead cheesemaking using the best and freshest milk.

We made a decision to keep our cheese local to California to limit the greenhouse gas emissions of transporting out of state. Thankfully, we have an outstanding spokesperson for our cheese in Vivien Straus, who founded the Sonoma/Marin and California Cheese trail. Vivien is every small California cheesemaker's biggest cheerleader. She volunteers her time on the maps and apps and speaking to publications for one reason: to ensure that we all succeed and keep making cheese.

Each state has its own amazing cheeses, and we encourage you to eat what is local, whether that be cheese, fruit, veggies, or meat.

It takes many hands and ideas to create these cheeses—everyone brings their own experiences to the table to create extraordinary cheese.

Our Cheeses

Many of the recipes in this cookbook are made with our cheeses. Much like any seasonal ingredient, like veggies or fruit, we don't make all of our cheeses all year round; rather, each is a delicious reflection of its season. Cheesemakers Nick and Oscar have really pushed the concept of sticking to the seasonality of our cheeses and not giving in to the pressure of what the consumer may want at any given time. In the winter and early spring, when the goats and sheep have yet to kid and lamb, we don't have a great deal of milk, so the cheeses we offer are aged: Atika, El Greco, and Koto'la. During late spring and summer, when milk is plentiful, we can offer our fresh cheeses: Liwa, Teleeka, Bossy, and Kenne.

The following is a guide to the flavors you'll find in our cheeses that we reference throughout, so you can find a local version that is close to what we're using:

LIWA (FRESH GOAT)
Season: Made year-round, best in spring
Tasting notes: Tart, fresh, bright
Pair with: Early strawberries, crusty bread, local honey

KENNE (SOFT-RIPENED GOAT)
Season: Early summer
Tasting notes: Tangy, earthy, creamy
Pair with: New potatoes, dill, snow peas

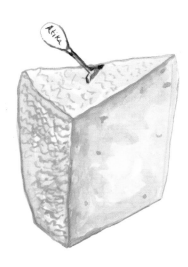

TELEEKA (SOFT-RIPENED GOAT, SHEEP, AND COW)
Season: Made year-round, best in early summer
Tasting notes: Buttery, citrusy, creamy
Pair with: Dark chocolate, roasted winter squash, sliced radishes

ATIKA (AGED GOAT AND SHEEP)
Season: Made in summer, available at different ages year-round
Tasting notes: Luscious, smooth, and nutty when younger; piquant and more textured when further aged
Pair with: Roasted ears of corn, blackberries, brined olives

KOTO'LA (BRINED GOAT)
Season: Made in late summer and autumn, available year-round
Tasting notes: Briny, tart
Pair with: Apricots, garden tomatoes, fresh red chiles

BOSSY (SOFT-RIPENED COW)
Season: Late fall and winter
Tasting notes: Mushroomy, buttery, creamy
Pair with: Chanterelles, baked apples, toasted challah

EL GRECO (BRINED COW)
Season: Winter
Tasting notes: Briny, buttery, tangy
Pair with: Sliced pear, mandarins, slow-cooked beans

A Recipe FOR Renewal

BY JESSICA MACLEOD

In the ebb and flow of life, there are moments that stand out, seemingly insignificant yet profoundly transformative—inviting you to come back home to your true nature. This is the story of how my part in this cookbook came to be, woven from threads of resilience, connection, and the simple joy of learning to be creative again.

My family has been in California for five generations. We moved around quite a bit when I was a kid, and I was always fed amazing home-cooked meals no matter where home was. I was raised by a single mom who was never afraid to experiment in the kitchen and go on the search for obscure ingredients. I also spent a lot of time with my grandparents—they had a big influence on my childhood and all had their own signature dishes and cooking styles, many of which are recreated in the chapters of this book.

We settled in Marin when I was in seventh grade, and I feel lucky to call it home: misty redwood forests, sage-scented mountains, flowing creeks, wild beaches, and nearly 60,000 acres of protected agricultural land—mostly worked by small-scale organic farmers. Turkeys and deer roam the neighborhoods, bald eagles can be spotted, and whales spout off the coast. This place is deeply magical, and I'm so grateful for the conservationists who have put decades of work into protecting it for future generations to enjoy.

Although my heart lies here, I graduated from college during the financial crisis and felt compelled to move across the country to Washington, DC, and build a serious career in public policy and technology. While I eventually made it back to Marin without a plan and only a couple of suitcases, something was still missing. After fifteen years in front of a computer screen, my escape fantasies became less and less about making it big and climbing the ladder in my field—they were about finding a way to live more simply and more connected to the natural world.

I began to find refuge in quiet moments in the kitchen, visiting the farmers' markets, learning about our native edible flora, and recreating traditional dishes using our local ingredients. With encouragement from my husband, Doug, I reached out cold to Toluma Farms asking if they would ever entertain a volunteer. I started by offering my technical and management skills, but happily, Tamara replied instantly inviting me to spend the day shadowing the herd manager and assisting in the barn with some pregnant goats.

I returned week after week, hauling hay bales around, supporting mother goats in labor, and bottle-feeding their kids. These long, quiet days in West Marin were some of the most peaceful I had had in years and left me with a profound sense of joy that I knew I needed to follow. With wonderful cheese to take home each week and a quick stop at the farm stands, I began recipe testing, jotting down what I made, and painting the ingredients in a family recipe book. The sweetness of this ritual provided a powerful antidote to long hours in corporate tech.

I've come to believe that the best ingredient for life is well-harnessed enthusiasm. So, I approached Tamara one afternoon at the end of my volunteer shift and told her about the family cookbook featuring so much of her cheese. She too had recipes, and the farm staff had recipes that could be shared and celebrated. We also had stories to tell and philosophies to share about community, climate change, and our choice to stay optimistic about the future.

In 2022, with a newborn in tow, I embarked on the journey of recipe testing, drawing inspiration from old cookbooks and the wisdom of seasoned chefs that have sourced Tomales Farmstead Creamery cheeses for years. As the pages of our cookbook took shape, so too did a new chapter in my life. Filled with a renewed sense of purpose and gratitude, I found myself embracing a slower, more intentional way of living, guided by the rhythms of the seasons and the wisdom of generations past.

Now, as I share this collection of recipes and reflections with the world, I am filled with a deep sense of gratitude for the journey that brought me here. May these pages serve as a reminder to cherish the simple moments, to savor the flavors of life, and to honor the interconnectedness of all living things. I have found a new commitment in the creation of this book, to follow the immortal instructions of Mary Oliver: "Pay attention. Be astonished. Tell about it."

SPRING

New Life: Kidding, Lambing, and Green, Green Grass

Our family has been fortunate enough to travel to many places throughout the world, and we are absolutely convinced that there is no place more magical than coastal Marin in springtime. Once you've experienced it—with hundreds of baby goats and lambs running around in damp, fragrant fields—it's the only place you'll ever want to be. We never take this beauty for granted, and we do all we can to protect every blade of grass, drop of free-flowing water, and inch of soil so this slice of natural landscape can flourish long into the future.

David and I are first-time farmers, meaning we are the first after a gap of several generations to farm professionally. My grandparents (and many great-grandparents) once farmed in rural Oklahoma and New York, starting in the 1700s, but we ourselves were not raised on farms; we pursued professions in health care (me as a clinical psychologist, and David as a scientist-surgeon). Our professional focus has always been supporting people to be their healthiest selves, both emotionally and physically, so in many ways the allure of farming was a natural for us. My trainer and friend (Kelly Redanz) has taught me how to be physically stronger than I thought possible. In our twice-a-week sessions, we discuss how human wellness and the health of our environment are inextricably linked. Our collective resilience, well-being, nutrition, and ability to avert diseases is undoubtedly connected to the food we eat, the water we drink, and the air we breathe.

There are mountains of evidence that an urban walk or a day spent hiking has a host of benefits, including improved attention span, lower stress, better mood, and even upticks in empathy and cooperation. While most research so far has focused on green spaces like parks and forests, researchers are now also turning to the benefits of *blue* spaces, such as places with river or ocean views. It is encouraging to see scientists and medical providers charting a new course for policymakers and the public to better harness the healing powers of Mother Nature.

Then there are the animals. On more occasions than I could have ever anticipated, someone visiting us will hold a baby goat in their arms and begin to cry. Sometimes it's a grown man, surprised at their own overwhelming burst of emotion. Sometimes it's a tech worker who hasn't allowed herself to slow down in nature with animals vying for love. Sometimes it's a young child who may leave inspired to pursue farming one day.

TO START
TO START
TO START
TO START
TO START

TO START
TO START
TO START
TO START
TO START

Wild Mushroom, Thyme, and Liwa Toasts

In Marin, our mushroom season begins after the first rainfall of autumn and lasts into early spring. The best times to forage are usually in January, about one week after a rainstorm that dropped 2 inches [5 cm] or more of water. There are incredible regions to forage along the coast, in the mountains, and even in the less traveled areas of town where pine, Douglas fir, and oak trees grow.

Unless you're an expert yourself, always forage with a local guide or group of experts (like our Mycological Society of Marin County) to ensure you're able to accurately identify edible and delicious mushrooms while avoiding the toxic varieties. If you're unable to forage with an expert, visit your local farmers' market in the early spring and find morels, chanterelles, trumpets, porcini, and more.

MAKES 12 SERVINGS

12 ¼ in [6 mm] slices of fresh sourdough baguette

2 Tbsp extra-virgin olive oil (we love McEvoy Ranch Olive Oil)

3 Tbsp unsalted butter

2 Tbsp chopped shallots

1 tsp chopped garlic

2 cups [200 g] finely chopped wild mushrooms

½ tsp kosher salt, plus more as needed

2 Tbsp chicken stock

8 oz [230 g] fresh goat cheese (like our Liwa)

2 Tbsp chopped fresh thyme leaves, plus twelve 2 in [5 cm] sprigs, for garnish

Freshly ground black pepper

1. Preheat the oven to 350°F [180°C].

2. Arrange the baguette slices on a baking sheet and brush with the olive oil. Bake for about 5 minutes or until lightly golden and slightly crisp. Let cool on the baking sheet.

3. In a medium skillet or sauté pan over medium heat, add 2 Tbsp of the butter and cook until the foaming subsides and the butter begins to brown. Add the shallots and garlic and sauté for 2 to 3 minutes until softened. Add the mushrooms and salt and sauté until soft, about 8 minutes. Add the chicken stock, scrape the brown bits from the bottom of the pan, and cook for 1 minute to reduce the liquid. Add the remaining 1 Tbsp of butter and 2 oz [55 g] of the cheese, remove from the heat, and stir to melt the cheese. Fold in the chopped thyme and season with salt and pepper.

4. Spread the remaining cheese evenly across the baguette slices. Spoon about 1 Tbsp of the hot mushrooms onto a baguette slice. Repeat with the remaining baguette slices. Garnish with thyme sprigs and serve immediately. We recommend eating this dish the same day it's made.

Peas, Cheese, and Caviar!

This recipe is contributed by Seth Stowaway, owner and chef at Osito.

Chef Seth Stowaway of Osito in San Francisco is all about combining playfulness with elegance, and this recipe does just that. Seth and his team cooked a fabulous summer dinner on the farm and this delicious appetizer, with healthy portions of Tsar Nicoulai caviar, was a huge hit. Tomales Farmstead Creamery has partnered with Tsar Nicoulai for a few tasty events as our fresh chèvre pairs perfectly with their caviar, and we have similar regenerative farming practices.

MAKES 8 TO 10 SERVINGS

¼ cup [55 g] unsalted butter

1 qt [480 g] double-shucked English peas

2 cups [480 ml] heavy cream

8 oz [230 g] fresh goat cheese (like our Liwa), at room temperature

Kosher salt

2 oz [55 g] caviar (we like Tsar Nicoulai Estate Caviar)

1. In a 2 qt [1.9 L] pot over low heat, melt the butter gently (do not brown), then fold in the peas. Add the cream and cook over low heat until the peas are tender but still bright green. Fold in the room-temperature cheese until fully incorporated, then remove from the heat.

2. Place the mixture in a blender and blend until smooth. Pass the blended mixture through a fine-mesh sieve into a medium mixing bowl. Add salt to taste.

3. Fill a large bowl with ice and place the bowl with the mixture directly on top. Spin the top bowl using a spatula to move the mixture, allowing it to cool quickly and evenly.

4. Add 2 cups [180 g] of the mixture to a high-powered blender and, starting at medium-low speed, begin mixing the ingredients, gradually increasing to a high speed. Whip the mixture for 2 minutes, or until light and fluffy, resembling a mousse or thick whipped cream. Alternatively, if you have a Thermo Whip dispenser, add 2 cups [180 g] of the mixture to it and charge twice, then turn it upside down and give it a good shake. Repeat with extra mixture as needed.

5. In a small serving bowl, spoon a small portion of caviar at the bottom. Gently top with the goat cheese and pea mousse and serve immediately. This dish can be stored in the refrigerator in an airtight container for a couple of days.

Asparagus and Teleeka Tart

We always feel like we're cheating when we use store-bought puff pastry, but let us tell you, it's the right thing to do in this case. This tart can be cut into small squares and served as an hors d'oeuvre with cocktails, cut into large squares and paired with a salad for an elegant lunch, or portioned as a first course. We recommend using very thin asparagus sliced in 1 in [2.5 cm] pieces and putting your artistic side to work to build beautiful patterns and designs atop the pastry. Not only is it fun, but it prevents anyone from getting a huge string of asparagus that requires multiple bites. Lastly, we use three Tomales Farmstead Creamery cheeses in this recipe—Teleeka, Liwa, and Atika—so this is truly a celebration of the cheese!

MAKES 6 TO 8 SERVINGS

1 sheet (12 oz [340 g]) all-butter puff pastry, thawed

6 oz [170 g] fresh goat cheese (like our Liwa), at room temperature

6 oz [170 g] triple cream cheese (like our Teleeka), at room temperature

¼ cup [60 ml] whole milk

2 garlic cloves, minced

1 Tbsp finely grated lemon zest

½ tsp kosher salt

½ tsp freshly ground black pepper, plus more for topping

1 large egg, at room temperature

25 to 30 extra-thin asparagus, woody ends removed, cut into 1 in [2.5 cm] pieces

1½ oz Parmesan-style cheese (like our Atika), shaved

1. Preheat the oven to 425°F [220°C].

2. Gently roll out the puff pastry to a 13 by 10 in [33 by 25 cm] rectangle, then place it on a lined baking sheet. Using a paring knife, score a 1 in [2.5 cm] border and use a fork to prick all around the inside of the border to prevent puffing in the middle. Blind bake the pastry for 5 to 7 minutes until lightly golden and slightly puffed. Remove and let cool, pressing down the center and leaving the border puffed.

3. In a medium bowl, combine the fresh goat and triple cream cheeses, milk, garlic, lemon zest, salt, and pepper. Whisk the egg and add it into the cheese mixture to combine.

4. Spread the cheese mixture evenly in the center of the prepared pastry. Arrange the asparagus in a pattern of your choice to cover the entire center of the pastry. Top evenly with the shaved Atika.

5. Bake until the asparagus is cooked and the cheese is melted, 10 to 15 minutes. Season with more black pepper.

6. Remove the tart from the oven and let it cool to warm or room temperature before slicing and serving. The tart can be stored in an airtight container in the refrigerator for up to 3 days and reheated in the oven.

Spicy Kale and Avocado Salad with Farro and Koto'la

Kale is so good for you, and yet it can be so tough to eat! The trick with this recipe is to massage the kale with your hands and allow the acid in the dressing to break down the fibrous leaves, transforming them into a velvety, flavorful treat. Served with a chewy and filling pile of farro and dotted with a salty bit of Koto'la, you have a perfect lunch that is quick and easy. We recommend making this dish with liquid aminos—but if you can't find it, substitute low-sodium soy sauce.

MAKES 4 SERVINGS

- 1 head (8 to 10 oz [230 to 280 g]) curly green kale
- ¼ cup [60 ml] extra-virgin olive oil
- 1 small lemon, juiced (about 2 Tbsp)
- 1½ tsp Bragg Liquid Aminos or low-sodium soy sauce
- 1 tsp sriracha
- 1 cup [160 g] cooked farro, al dente
- 1 ripe avocado, diced
- 3 oz [85 g] feta-style cheese (like our Koto'la), crumbled

1. Cut the kale leaves away from the stalks and tear them into roughly 2 in [5 cm] pieces (discard the stalks). Place in a large salad bowl.

2. In a separate bowl, with a fork or small whisk, whisk together the olive oil, lemon juice, liquid aminos, and sriracha. Pour enough dressing over the kale to lightly coat the leaves, then use your hands to massage the dressing into the leaves—this will break down the toughness of the kale and allow it to soften. Refrigerate the kale for about 15 minutes.

3. To serve, plate about ¼ cup [40 g] of cooked farro per person, spreading it on the plate from the center outward to create a well in the center. Add about one handful of kale to the center of each plate. Top with the avocado and crumbled cheese. We recommend eating this dish the same day it's made—it can be stored in the refrigerator for up to 8 hours.

Endive, Pear, and Kenne Salad with Toasted Walnuts and Honey

We love cheese boards and crudité boards loaded with fresh produce. A little bit retro and elevated, this dish combines the best of both in individual bite-sized portions. Best of all, you can make it in about 10 minutes and put it out for a party. Get ready for compliments.

MAKES 6 TO 8 SERVINGS

2 heads green endive, leaves separated

1 head red endive, leaves separated

2 ripe Bartlett pears, thinly sliced

8 oz [230 g] soft-ripened goat cheese (like our Kenne), crumbled

1 cup [120 g] chopped walnuts, lightly toasted

¼ cup [85 g] honey, slightly warm and runny

1. Arrange the endive leaves in slightly overlapping layers on a platter, cup side up.

2. Place a pear slice or two (depending on the size of the leaf) into each leaf. Top with the cheese and walnuts. Drizzle honey lightly across all the leaves.
    We recommend eating this dish the same day it's made—it can be stored in the refrigerator for up to 4 hours with the honey drizzled just before serving.

The Farm HEALED ME

BY JULIA BRANDT, FORMER APPRENTICE AND CURRENT SHEEP AND WORKING DOG MANAGER

Two years before arriving to Toluma Farms, I was living in Southern California and taking a holistic approach to managing my polycystic ovarian syndrome (PCOS) symptoms with no relief. Between an active lifestyle outside of my typical nine-to-five job and what seemed like a million supplements, I continued to struggle with high cortisol, prediabetic glucose levels, and a host of other hormonal imbalances.

In a leap of faith, I followed my gut when I came across an apprenticeship opportunity at Toluma Farms. My journey to the farm wasn't merely a change of scenery; it was a transformative shift in lifestyle that would profoundly impact my health and well-being. Little did I know that this move would hold the key to reversing the debilitating symptoms of my PCOS. As I immersed myself in the rhythm of farm life, something remarkable began to unfold. The demanding yet fulfilling work of tending to the land and animals became more than just a daily routine; it became a therapeutic regimen that rejuvenated my body and spirit. Gone were the days plagued by the relentless symptoms of PCOS. Instead, I found myself energized by the physical exertion of farm chores, and nourished by the wholesome foods that were readily accessible.

The stress-reducing effects of being surrounded by nature and engaging in meaningful work helped regulate my cortisol levels, which played a pivotal role in managing my PCOS symptoms. Three months into my apprenticeship, I received my bloodwork results with disbelief—my hormone levels and blood sugar had completely normalized.

It is at Toluma Farms that I discovered a profound connection between my health and the environment. The farm offered more than just sustenance; it provided a sanctuary for healing and renewal. With each passing day, as I toiled alongside the earth's gentle rhythms, I bore witness to the remarkable transformation taking place within me. PCOS, once a formidable adversary, now seemed like a distant memory as I embraced a life restored to balance on the farm.

Eggs Brouillés with Truffle Butter and Brioche

Scrambled eggs can be incredibly elegant when you dress them up for a special occasion. The pudding-like texture of the eggs in this recipe, paired with the back-palate notes of black truffle and the crisp but rich bite of brioche, take these eggs from your breakfast table straight to a dinner party.

MAKES 8 SERVINGS

One 14 to 16 oz [400 to 455 g] loaf of brioche bread

8 large eggs, at room temperature

1 Tbsp unsalted butter

¼ cup [60 g] crème fraîche

1 Tbsp black truffle butter

½ tsp kosher salt

2 oz [55 g] fresh goat cheese (like our Liwa), crumbled

4 fresh thyme sprigs, for garnish

Freshly ground black pepper

1. Preheat the oven to 350°F [180°C].

2. Slice the brioche into eight equal slices, each about 1½ in [4 cm] thick. Trim the crusts and cut the corners to create a circle or octagon shape. Using your thumb and forefinger, gently pinch the center of each brioche slice to create a 2 in [5 cm] crater. Place each slice on a baking sheet and lightly toast in the oven for 8 to 10 minutes, until golden and slightly crisp.

3. In a large bowl, whisk the eggs to break up the yolks. In a heavy saucepan over medium-low heat, melt the unsalted butter, stirring frequently, until slightly brown. Add the eggs and whisk vigorously in the pan for 30 seconds. Remove from the heat and whisk for another 30 seconds. Return to the heat and whisk again for 30 seconds. Repeat this step two more times until you begin to see the curd form and the egg mixture thicken. Take the egg mixture off the heat and whisk in the crème fraîche. Return to the heat and whisk in the truffle butter and salt. When the mixture resembles a thick pudding, remove from the heat.

4. Plate the brioche slices and spoon the egg mixture into the crater of each toast to form a mound. Top with a sprinkle of cheese and garnish with a half sprig of thyme. Season with pepper and serve immediately. We recommend eating this dish the same day it's made.

CHILDHOOD *on the Farm*

BY EMMY HICKS JABLONS,
TAMARA'S DAUGHTER AND
FARM PHOTOGRAPHER

My face is pressed against hot trampoline fabric. I play a familiar game of peeking through the cracks at the dirt, bugs, and shifting shadows beneath me. I lie here for a while. The sun heats up my back while I listen to eucalyptus trees swaying, the sound of feet shuffling in and out of puddles in the milking parlor. When I start jumping again, on a really good jump I can see miles past Middle Road. Past the horses, the glistening *estero*, the neighbor's building I always thought resembled a massive Morton salt shaker.

Down in the barn, pregnant does (goat mommas) named after famous pop stars like Rihanna and Katy Perry impatiently chew alfalfa and wait for their babies to come. The barn quickly fills with new voices that are unsure yet powerful enough to be heard from across the farm. As it begins to get dark, the kids settle into their first night in the barn. This is spring on our farm.

MAINS
MAINS
MAINS
MAINS
MAINS

MAINS
MAINS
MAINS
MAINS
MAINS

Farmer's Panzanella Salad

This recipe is dedicated to the small joy of dragging a crisp slice of sourdough along the bottom of the salad bowl to scoop up the remaining bits of dressing and greens. San Francisco and Marin are famous for authentic sourdough bread, and this salad has sourdough as its star.

This California twist on a traditional Italian panzanella is an all-green take on the classic that you can enjoy outside of tomato season and makes for the perfect farmer's lunch. We love fennel, cucumber, and asparagus in this salad, but you can substitute any crunchy fresh green vegetable that is in season. Vegetarians can substitute the pancetta with lightly toasted pine nuts, but if you're a meat eater, know that it adds a salty, rich bite.

MAKES 6 SERVINGS

SALAD

2 oz [55 g] pancetta, cut into ½ in [13 mm] pieces

6 cups [180 g] cubed sourdough bread, cut into 1 in [2.5 cm] cubes

1½ Tbsp extra-virgin olive oil

2 tsp kosher salt

1 tsp freshly ground black pepper

6 to 8 stalks thin asparagus, woody ends removed (about 5 oz [140 g])

1 fennel bulb, cut into 1 in [2.5 cm] pieces

1 English cucumber, diced into 1 in [2.5 cm] pieces

1½ cups [330 g] sugar snap peas, cut into 1 in [2.5 cm] pieces

10 to 15 fresh basil leaves, chopped

2 oz [55 g] pea shoots, chopped

2 oz [55 g] chopped fresh parsley

3 to 5 oz [85 to 140 g] feta-style cheese (like our Koto'la)

VINAIGRETTE

1 shallot, minced

2 tsp Champagne vinegar

½ tsp Dijon mustard

½ tsp freshly ground black pepper

1. Warm a dry sauté pan over medium-high heat, add the pancetta, and cook until crisp, about 3 minutes. Using a slotted spoon, transfer the pancetta to a plate and reserve the drippings for the vinaigrette.

2. To make the vinaigrette, in a large serving bowl, combine the shallot, 2 Tbsp of reserved pancetta fat (supplement with olive oil if needed), the Champagne vinegar, and mustard. Whisk to emulsify. Add the pepper and whisk again.

3. To make the salad, preheat the oven to 375°F [190°C].

4. In a large mixing bowl, toss the bread cubes with 1 Tbsp of the olive oil, 1 tsp of the salt, and ½ tsp of the pepper. Spread out the cubes on a baking sheet and toast in the oven until crisp and golden with lightly browned edges, 12 to 15 minutes.

5. Add the asparagus to a large mixing bowl. Toss with the remaining ½ Tbsp of olive oil, 1 tsp salt, and ½ tsp of pepper. Spread out the asparagus on a baking sheet and bake until lightly golden with browned edges, about 10 minutes. Cool, then dice into 1 in [2.5 cm] pieces and set aside.

6. Toss the prepared bread cubes in the vinaigrette, then add the asparagus, fennel, cucumber, and snap peas and toss. Add the basil, pea shoots, and parsley and toss gently. Crumble in the cheese and toss again before serving. We recommend eating this salad the same day it's made—it can be stored in the refrigerator for up to 1 hour.

Savory Leek, Artichoke, and Atika Tart

Leeks are one of our favorite vegetables, and we're always looking for new ways to celebrate their delicate flavor. Although there is nothing that quite compares to homemade pie crust and freshly steamed artichokes, if you're short on time you can achieve similar results using store-bought artichoke hearts cured in olive oil and puff pastry. Make sure to linger over the caramelization of the leeks, sautéing them over low heat and stirring occasionally to avoid sticking. Slowly caramelizing the leeks will result in a softer and slightly melty texture along with a greater depth of flavor and richness.

1. To make the tart dough, in the bowl of a food processor, add the flour, salt, and cold butter and pulse to form pea-size crumbs. With the motor running, add the ice water just until the dough comes together. Transfer the dough to a lightly floured work surface and shape it into a 1 in [2.5 cm] thick disk. Wrap the disk in plastic wrap and refrigerate for 30 minutes to 1 hour.

2. Preheat the oven to 400°F [200°C].

3. To make the filling, in a sauté pan over medium-low heat, warm the olive oil and butter. Add the leek and shallot and sauté until caramelized, 8 to 10 minutes. Set aside.

4. In a mixing bowl, combine the goat cheese, half of the lemon zest, the salt, and pepper. Add the leek mixture to the bowl along with the chopped artichoke hearts and toss to coat.

5. Line a baking sheet with parchment paper. On a lightly floured work surface, roll out the dough to a 12 in [30.5 cm] circle and transfer it to the prepared baking sheet.

6. Gently spread the cheese mixture over the dough, leaving a 1½ in [4 cm] border. Scatter the remaining lemon zest, the Parmesan-style cheese, and thyme on top. Fold the edge of the dough over the cheese filling. Lightly brush the pastry with the beaten egg and sprinkle with sea salt.

7. Bake on the center rack of the oven until the pastry is puffed and golden, 40 to 45 minutes. Transfer to a cutting board to cool slightly before slicing into wedges to serve. The tart can be stored in an airtight container in the refrigerator for up to 3 days.

MAKES 6 SERVINGS

TART DOUGH

1¼ cups [175 g] all-purpose flour, plus more for rolling

½ tsp kosher salt

½ cup plus 2 Tbsp [140 g] cold unsalted butter, cut into ½ in [13 mm] pieces

5 Tbsp [80 ml] ice-cold water

FILLING

3 Tbsp extra-virgin olive oil

2 Tbsp unsalted butter

1 large leek, white and light-green parts, thinly sliced

2 Tbsp minced shallot

8 oz [230 g] fresh goat cheese (like our Liwa)

Zest of 1 lemon

½ tsp kosher salt

½ tsp freshly ground black pepper

1 cup [220 g] drained, roughly chopped oil-cured artichoke hearts

8 oz [230 g] Parmesan-style cheese (like our Atika), finely grated

2 tsp fresh thyme leaves

1 large egg, beaten

Maldon sea salt

Liwa and Ham Frittata with Garden Lettuces

This simple and elegant brunch is quick to prepare and extremely nourishing. This is an extremely flexible recipe, so feel free to use any leftover vegetables, bacon, and cheese you have sitting in your fridge to ensure nothing goes to waste. During a recent stay at the farm, we made this recipe using duck eggs from our neighbors at True Grass Farms, a bunch of leftover baby spinach, and Molinari pancetta. By preparing your side salad while the frittata puffs in the oven, you'll have a beautiful brunch on the table in no time.

MAKES 6 SERVINGS

6 Tbsp [90 ml] extra-virgin olive oil

1 Tbsp salted butter

2 Tbsp minced shallot

8 large eggs

¼ cup [60 ml] heavy cream

½ tsp kosher salt

¼ tsp freshly ground black pepper

4 oz [115 g] fully cooked country ham or Canadian bacon, roughly diced

2 cups [40 g] baby spinach, stems removed

8 oz [230 g] fresh goat cheese (like our Liwa) or chèvre

2 oz [55 g] Parmesan-style cheese (like our Atika), finely grated

3 Tbsp Champagne vinegar

2 Tbsp coarse Dijon mustard

4 cups [80 g] mixed baby lettuces

1. Preheat the oven to 375°F [190°C].

2. In a large cast-iron or heavy ovenproof skillet over medium heat, warm 2 Tbsp of the olive oil and the butter. Add the shallot and sauté until translucent, stirring occasionally, about 5 minutes.

3. In a large bowl, whisk together the eggs, cream, salt, and pepper until frothy and set aside. Add the ham to the skillet and brown for 3 to 5 minutes. Once browned with slightly crisp edges, add the spinach and stir until slightly wilted, about 2 minutes. Add the egg mixture and slowly push the eggs around to evenly distribute the ham, spinach, and shallots throughout. Once large curds begin to form, dollop with large chunks of the fresh goat cheese.

4. Carefully transfer the skillet to the oven and bake until the eggs are set and the edges are starting to turn golden, about 15 minutes. Remove from the oven and sprinkle the Parmesan-style cheese over the top. Return the skillet to the oven to let the cheese melt, about 3 minutes.

5. Meanwhile, in a medium bowl, whisk together the remaining ¼ cup [60 ml] of olive oil, the vinegar, and mustard. Add the mixed baby lettuces to the bowl and toss in the dressing. Plate the salad.

6. Remove the frittata from the oven to cool until it starts to slightly pull away from the edges, about 10 minutes. Slice into wedges in the skillet, plate, and serve.

The frittata can be stored in an airtight container in the refrigerator for up to 3 days.

Spring Vegetable Risotto with Kenne, Atika, and Liwa

This spring vegetable risotto is the first recipe I, Jessica, ever made with cheese from Toluma Farms. After a long, misty spring day rebedding hay in the barn and bottle-feeding the (goat) kids, I was rewarded with an armful of cheese to take home. I stopped at Little Wing farm stand and Palace Market in Point Reyes to get some inspiration for what I could do with this bounty of cheese. Fresh asparagus, huge fragrant fennel bulbs, and a mountain of mushrooms caught my eye. These ingredients are fabulous in risotto, but the wonderful thing about this base is its versatility with any lovely in-season vegetables from your local farm stand or market—so get creative.

Yes, this is a lot of vegetables—but remember that this recipe is a celebration of cheese, added at the end of cooking in any exuberant quantity that delights you and your guests. The final dash of lemon, fresh herbs, and edible flowers balances the richness and ensures it feels as light as a spring day on the farm.

MAKES 6 SERVINGS

5 cups [1.2 L] chicken stock, preferably homemade

3 Tbsp extra-virgin olive oil, plus more for serving

3 Tbsp salted butter

1 leek, white and light green parts, chopped (about 1 cup)

2 cups [370 g] Arborio rice

½ cup [120 ml] dry vermouth

1 fennel bulb, chopped (fronds reserved for garnish)

Kosher salt

continued

1. In a small saucepan over medium-high heat, bring the chicken stock to a gentle simmer.

2. In a large stockpot over low heat, warm the olive oil and 2 Tbsp of the butter until lightly shimmering. Add the leek and sauté until translucent. Add the rice and stir to coat. Cook until it smells lightly toasted, about 2 to 3 minutes. Add the vermouth and scrape the bottom of the pan to loosen any sticky bits.

3. Add one ladleful of chicken stock at a time, stirring between additions and waiting until the previous stock is mostly absorbed into the rice before adding more. Adjust the heat to keep it at a low simmer until there is 1 cup [240 ml] of the stock left in the saucepan.

4. Add the chopped fennel and a pinch of salt. Stir and slowly add ¼ cup [60 ml] of the remaining stock until the fennel is tender, about 3 minutes.

continued

5. Gently stir the asparagus into the risotto, followed by the artichoke hearts. Add the remaining ¾ cup [180 ml] of stock and place the lid on while you cook the mushrooms.

6. In a medium sauté pan over medium-high heat, melt the remaining 1 Tbsp of butter, then add the mushrooms and a pinch of salt and sauté until reduced in size, aromatic, and tender, about 5 minutes. Stir the mushrooms into the risotto.

7. Once all remaining chicken stock has been absorbed, stir in the peas, lemon zest, and lemon juice. Stir in the cheeses, taste, and add salt and pepper as desired.

8. Serve in shallow bowls with a drizzle of good olive oil and additional cheese on top. Garnish with the fennel fronds, parsley sprigs, and edible flowers. This dish can be stored in an airtight container in the refrigerator for up to 3 days.

15 to 20 stalks thin asparagus, woody ends removed, cut into 1 in [2.5 cm] pieces

6 baby artichoke hearts, cut into 1 in [2.5 cm] pieces

3 cups [180 g] mix of fresh mushrooms (like cremini, oyster, and porcini), roughly chopped

½ cup [60 g] fresh or frozen peas

Zest and juice of 1 lemon

¼ cup [55 g] soft-ripened goat cheese (like our Kenne), plus more for serving

3¼ oz [90 g] Parmesan-style cheese (like our Atika), grated, plus more for serving

¼ cup [70 g] fresh goat cheese (like our Liwa) or fresh chèvre, plus more for serving

Freshly ground black pepper

2 to 3 fresh parsley sprigs

½ cup [6 g] edible flowers to garnish

Roast Chicken with Fennel and Olives

This is a Sunday night staple for our family. All you need to do is a bit of light chopping in the afternoon, the chicken goes into the oven at 5 p.m., and dinner is on the table by 6:45 with plenty of leftovers for the week. And always save the carcass for making stock. The vegetables caramelize in the chicken drippings and become incredibly sweet, while the olives add a salty and creamy bite. We recommend serving with lots of extra sourdough bread for scooping up the roasting pan drippings.

MAKES 4 SERVINGS

2 fennel bulbs, diced

1½ cups [355 g] Castelvetrano olives

2 large carrots, diced

½ yellow onion, diced

10 to 12 garlic cloves, skins left on

2 Tbsp extra-virgin olive oil

6 lb [2.7 kg] whole organic roasting chicken

1 lemon, halved

10 to 15 fresh thyme sprigs

4 to 5 fresh rosemary sprigs

3 Tbsp salted butter, melted

1½ Tbsp kosher salt

6 to 10 fresh parsley sprigs, stems removed and leaves roughly chopped

Freshly ground black pepper, to serve

1. Preheat the oven to 425°F [220°C].

2. Arrange the fennel, olives, carrots, onion, and garlic in the bottom of a large roasting pan or cast-iron skillet. Drizzle with the olive oil and toss to lightly coat.

3. Remove the giblets if they've been left in the chicken cavity, pat the chicken dry, and stuff with the lemon, thyme, and rosemary. Truss the chicken: Place the chicken in the pan, breast-side up, on top of the vegetables, and cross the legs, tying them tightly with cotton kitchen twine. Alternatively, you can create a natural truss by puncturing a small hole in the skin of one thigh with a paring knife and slotting the opposite leg bone through it. Drizzle the chicken with the melted butter and sprinkle generously with the salt.

4. Roast the chicken for 90 minutes, turning halfway through. The skin should be slightly blistered and crisp across the top of the chicken, and the thickest part of the thigh should read 165°F [75°C] on a meat thermometer. Remove the chicken from the pan, turning it vertically to release any captured juices into the pan. Place it on a wooden cutting board and tent with aluminum foil.

5. Return the pan to the stovetop over medium heat and gently scrape the bottom with a wooden spoon to release any stuck-on bits. Reduce the pan drippings by about one-third to concentrate the flavor, about 10 minutes.

6. When ready to serve, carve the chicken and arrange with the vegetables on a platter. Pour the drippings over the chicken. Garnish with the parsley and season with pepper to serve. Leftovers can be stored in an airtight container in the refrigerator for up to 5 days.

Marinated Lamb Chops

Lamb chops are an incredibly fast main course and feel almost too fancy for a weeknight, but trust me, they're not. They're beautiful to present on a platter, absolutely tender and delicious thanks to a yogurt marinade, and filled with aromatic delight.

MAKES 4 SERVINGS

1 cup [240 g] whole-milk Greek yogurt

¼ cup [60 ml] extra-virgin olive oil

4 tsp minced garlic

1 tsp ground cumin

1 tsp ground coriander

Zest and juice of 1 lemon

2 lb [910 g] lamb chops, about 1½ in [4 cm] thick, separated from the rack and patted dry

1½ tsp kosher salt

1 tsp freshly ground black pepper

2 fresh rosemary sprigs

3 to 4 Tbsp roughly chopped fresh parsley, for garnish

1. In a small bowl, stir together the yogurt, 2 Tbsp of the olive oil, the garlic, cumin, coriander, lemon zest, and half of the lemon juice.

2. Brush the remaining half of the lemon juice over the lamb chops. Season the lamb with the salt and pepper and place in a zip-top bag for marinating. Add in the yogurt mixture and rosemary sprigs, turn to coat the lamb, and refrigerate for 30 minutes or up to 1 hour.

3. Remove the lamb from the marinade and remove the excess yogurt mixture from each lamb chop with a spatula so that the meat is peeking through. (If the yogurt coating is too thick, the lamb will likely burn before the lamb chop is cooked through.) Discard the remaining marinade and rosemary sprigs.

4. In a large sauté pan over medium-high heat, warm the remaining 2 Tbsp of olive oil. Working in two batches, cook the lamb until crusty and browned, 3 to 4 minutes per side. The yogurt in the marinade will help them take on color quickly.

5. To serve, arrange the cooked chops on a serving platter and garnish with the parsley. The cooked lamb chops can be stored in an airtight container in the refrigerator for up to 2 days.

Slow-Braised Lamb Shanks with Apricot Couscous

Gigot à la cuillère, or "lamb you can eat with a spoon," is an old-world dish from Languedoc, France, in which lamb shanks are braised for several hours in a brothy vegetable stew with cognac and herbs. However, many versions of this dish are extremely popular wherever in the world you see flocks of sheep—including at Toluma Farms!

We once tripled this recipe for a large dinner party at the farm, using quartered lamb legs. However, when serving a smaller crowd, we recommend ½ lb [230 g] of lamb per person and using lamb shanks rather than legs, which will give you more control over the portion sizes. While lamb is freshest in spring, this is a lovely warming dish to serve in the fall or winter and can be made with previously frozen lamb shanks.

There is very little hands-on cooking time, but it does require nearly the full day in a low oven, so stay close by, linger near the kitchen, and enjoy the ever-intensifying aromas as the meat falls off the bone. Get your spoons ready!

MAKES 6 SERVINGS

LAMB STEW

2 cups plus 6 Tbsp [480 ml plus 90 ml] extra-virgin olive oil

2 garlic heads, halved horizontally

10 fresh rosemary sprigs

3 Tbsp kosher salt

2 Tbsp freshly ground black pepper

3 lb [1.4 kg] lamb (3 bone-in shanks, or 1 leg, quartered)

1 Tbsp coriander seeds

8 oz [230 g] lamb sausage, casings removed

1 yellow onion, roughly diced

1 Roma tomato, roughly diced

1 lb [455 g] carrots, roughly diced

1 fennel bulb, roughly diced

1 turnip, roughly diced

3 garlic cloves, minced

10 fresh thyme sprigs

1 Tbsp ground cumin

1 tsp sweet paprika

½ cup [120 ml] cognac or Armagnac

6 cups [1.4 L] chicken stock

continued

1. To make the lamb stew, in an airtight container or zip-top bag, combine 2 cups [480 ml] of the olive oil, the garlic, rosemary, salt, and pepper. Add the lamb shanks and marinate in the refrigerator overnight or for up to 3 days. When you're ready to cook, remove the lamb from the marinade, pat lightly with a paper towel to dry, and let it sit for 30 minutes (1 hour maximum) to bring it to room temperature.

2. Preheat the oven to 300°F [150°C].

3. Heat a large cast-iron braising pot or rondeau over medium heat. Toast the coriander seeds for 30 seconds or until fragrant. Remove from the pot and lightly crush them with a mortar and pestle or the back of a cold pan. Set aside. In the same pan over high heat, add 2 Tbsp of the olive oil and brown the lamb shanks in batches, turning every 4 to 5 minutes until a deep brown crust forms on all sides of the shanks. Add the sausage to the pot and break it up with a wooden spoon. Let brown, only stirring occasionally, about 8 to 10 minutes. Set each piece aside on a platter after browning.

4. Once all the meat is browned, lower the heat to medium. Add the remaining ¼ cup [60 ml] of olive oil to the pot and add the onion, tomato, carrots, fennel, and turnip, scraping the bottom and stirring occasionally. Add the minced garlic, thyme, cumin, toasted coriander seeds, and paprika, stirring to coat the vegetables. Cover and let steam for 2 to 3 minutes.

continued

5. Add the cognac to the pot, scraping the bottom to release the stuck-on bits. Add the meat back to the pot along with the chicken stock. Make sure the meat is mostly submerged. Add water to cover if needed.

6. Cover the pot and increase the heat to high to bring to a simmer. Once simmering, transfer the pot to the oven and braise for 6 to 7 hours. Pull the pot out every 2 hours to gently stir and check on the consistency of the meat. When the bones from the shanks are loose in the pot and can easily be removed with tongs, it's done. Remove the thyme sprigs, then leave the pot on the stovetop with the lid on to keep warm.

7. To make the couscous, in a medium saucepan over medium heat, melt the butter. Add the shallot and stir until softened, 3 to 5 minutes. Add the chicken stock and bring to a boil. As soon as it comes to a boil, remove the pan from the heat and stir in the couscous and apricots. Cover and let it sit for 10 minutes until all the liquid is absorbed. Using a fork, fluff the couscous. Add the toasted pine nuts and three quarters of the parsley and fold very gently to incorporate.

8. To serve, spoon a large helping of couscous in the center of each plate, making a crater in the center and pushing outward on the plate to form a ring shape. Spoon the lamb and vegetables into the center of the ring. Top with a sprinkle of the remaining parsley and large crumbles of cheese. The lamb can be stored in an airtight container in the refrigerator for up to 5 days.

COUSCOUS

2 Tbsp salted butter

1 shallot, minced

6 cups [1.4 L] chicken stock

3 cups [420 g] couscous

6 oz [170 g] dried apricots, chopped

½ cup [60 g] pine nuts, toasted

1 bunch fresh parsley, leaves roughly chopped

4 oz [115 g] feta-style cheese (like our Koto'la), crumbled

NEXT *Gen*

BY SARAH ROLAND, FORMER APPRENTICE, OWNER OF BAYOU SARAH FARMS IN LOUISIANA

I knew I had to start a farm. Mainly because I love working outside, but also because I have family land, coupled with the belief that if we want to eat whole, natural, and fresh food, we in the local communities are responsible for producing our own. I started as a chicken farmer because I had the setup and knew birds. Also, dairy required more infrastructure (fences/pastures) than chickens. A few seasons of twelve-hour days, not to mention the poop water spills into my boots, confirmed that I am not interested in raising chickens for meat. My work with the chickens did help get my name out there in the local community, which is important for any farmer selling food. Chickens also gave me a foot in the door at the farmers' markets—since I was the only one selling chicken—and I started selling blueberries and honey that I had harvested.

So I returned to my original goal—a goat dairy. I apprenticed on a couple of farms, including Toluma Farms. There I developed the confidence that I could return to Louisiana and start my own dairy, but with buffalo—water buffalo, that is! It turns out that buffalo are perfect for a Louisiana climate: They are self-sufficient (like me), resistant to most parasites (unlike goats), and not susceptible to mastitis or hoof problems (again, unlike goats). Their milk has three times the butterfat compared to cow milk, and they thrive on a 100 percent grass-fed diet.

For me, buffalo farming is the most rewarding lifestyle I could possibly think of, and I am so grateful to be living it!

SWEET
SWEET
SWEET
SWEET
SWEET

SWEET
SWEET
SWEET
SWEET
SWEET

Crisp Popovers with Goat Kefir and Mulberry Rhubarb Compote

The popover is an American version of Yorkshire pudding and similar batter puddings made in England since the seventeenth century. There are good reasons for the staying power of the popover: They are unbelievably crisp and savory on the outside while maintaining a lighter-than-air texture and hint of sweetness on the inside. Popovers can be served plain in a big cloth-lined basket at the center of the table with brunch or dinner. They can also be adorned with lots of butter, sauces, and condiments, as we do in this recipe, for a not-too-sweet but satisfying dessert.

MAKES 12 SERVINGS

RHUBARB COMPOTE

3 rhubarb stalks (about 14 oz [400 g]), cut into ½ in [13 mm] pieces

2 cups [310 g] mulberries or raspberries

3 Tbsp granulated sugar

Zest of 1 lemon (about 2 Tbsp)

POPOVERS

4 large eggs, at room temperature

1½ cups [360 ml] whole milk, at room temperature

1½ cups [180 g] sifted all-purpose flour

¾ tsp fine salt

3 Tbsp unsalted butter, melted and cooled

Goat kefir or whole milk yogurt, for serving

1. Preheat the oven to 425°F [220°C].

2. To make the rhubarb compote, in a saucepan over medium heat, add the rhubarb, mulberries, ¼ cup [60 ml] of water, the sugar, and lemon zest and stir. The rhubarb mixture will bubble and thicken, 7 to 10 minutes. Lower the heat to low and cover for 25 minutes, stirring occasionally. When the rhubarb mixture has a thick jam-like texture, remove from the heat and pour it into a serving carafe to cool.

3. To make the popovers, grease a 12-cup muffin tin with vegetable oil and place in the oven for 5 minutes to preheat. In a blender, combine the eggs, milk, flour, and salt. With the blender running on low speed, slowly add the melted butter until just combined. The batter will be very thin.

4. Pour the mixture evenly into each cup of the muffin tin, filling about halfway up the cup. Bake for 30 minutes without opening the oven door. The popovers will grow substantially over the top of the cups. When fully cooked, gently transfer the popovers from the muffin tin to a cloth-lined serving basket.

5. To serve, place the carafe of rhubarb compote on the table along with a carafe of goat kefir. The popovers can be kept in an airtight container or bag at room temperature for up to 2 days.

Meyer Lemon and Rosemary Panna Cotta

Panna cotta is a refreshingly light and sweet dessert to serve as the weather starts warming up and you can't be bothered to turn on the oven. It's a flexible dish that can be made well in advance, prepared for a group or just for you, and can be flavored in endlessly creative ways. Meyer lemons and rosemary are incredibly abundant in the Mediterranean climate of Northern California, and this is a lovely way to celebrate the season and the perfume of the garden right on your plate.

MAKES 4 SERVINGS

⅛ oz [4 g] unflavored gelatin sheets or unflavored powdered gelatin

¼ cup [60 ml] whole milk

1 cup [240 ml] heavy cream

2 fresh rosemary sprigs

½ cup [120 ml] Meyer lemon juice, from about 3 lemons

¼ cup [50 g] granulated sugar

Thin lemon slices and edible flowers, such as borage or pansies, for garnish

1. In a 2 qt [1.9 L] saucepan, sprinkle the gelatin over the milk and heavy cream. Let the mixture stand for 2 minutes. Place the saucepan over medium heat and cook, stirring constantly, until the gelatin is melted. Add the rosemary sprigs and remove from the heat to steep the rosemary for about 10 minutes.

2. Remove the rosemary, and add the lemon juice and sugar and whisk to combine.

3. Pour the panna cotta mixture into four teacups or cocktail glasses until the liquid nearly reaches the top of the glass. Transfer to the refrigerator to begin to set, 2 to 3 hours.

4. To serve, add 1 very thin lemon slice to the top of each glass along with an edible flower such as borage or a pansy. The panna cotta can be stored in an airtight container in the refrigerator overnight.

Herbaceous Cocktail Cookies

These cookies fall somewhere between savory and sweet. They are delicately flavored with a variety of garden herbs, making them perfect for pairing with cocktails as an hors d'oeuvre. You and your guests will be sneaking back into the kitchen for a few more after dinner—guaranteed.

MAKES 24 COOKIES

2 cups [440 g] unsalted butter, at room temperature

1 cup [200 g] granulated sugar or superfine sugar

3 Tbsp roughly chopped fresh lavender flowers

3 Tbsp finely chopped fresh rosemary leaves

2 Tbsp finely chopped fresh thyme leaves

2 tsp vanilla extract

¼ tsp kosher salt

4 cups [560 g] all-purpose flour, plus more for dusting

1. Preheat the oven to 350°F [180°C] and line two baking sheets with parchment paper.

2. In a stand mixer or with a hand mixer, cream the butter and sugar on medium-high speed until pale and fluffy, about 3 minutes. Add the lavender, rosemary, thyme, vanilla, and salt. Mix until combined and the herbs are evenly distributed. With the mixer running on low speed, gradually add the flour 1 cup [140 g] at a time until fully combined.

3. Turn out the dough onto a clean, lightly floured work surface. Roll out the dough to about ⅛ in [4 mm] thick. Flour a roughly 2 in [5 cm] in diameter cookie cutter in the shape of your choice (I use rounds or hearts) and stamp out each cookie, placing them on the baking sheets about 2 in [5 cm] apart.

4. Bake for 8 to 10 minutes until the edges of the cookies are slightly golden but not browned. Let the cookies cool on the baking sheet for 5 minutes, then transfer to a wire rack to finish cooling.

   The cookies can be stored in an airtight container at room temperature for up to 1 week.

Great-Grandma Opal's Chocolate Sheet Cake

Tamara lived in northeast Oklahoma from a couple of weeks old till third grade and has many great memories of time spent on her great-grandparents' (Opal and Joe's) homestead. Opal would make her chocolate sheet cake for the Happy Hill Church gatherings—this classic cake was always such a crowd-pleaser. Tamara's mother made the cake for birthday celebrations, and Tamara continued that tradition with her own family.

MAKES 15 SERVINGS

CAKE

2 cups [280 g] flour

2 cups [400 g] granulated sugar

2 tsp ground cinnamon

1 tsp baking soda

½ cup [110 g] unsalted butter, at room temperature, plus more for greasing

½ cup [110 g] lard, at room temperature

¼ cup [20 g] cocoa powder

2 large eggs, at room temperature, beaten

½ cup [120 ml] buttermilk

1 tsp vanilla extract

ICING

½ cup [110 g] unsalted butter

6 Tbsp [90 ml] whole milk

¼ cup [20 g] cocoa powder

3½ cups [420 g] confectioners' sugar

1 tsp vanilla extract

1. Preheat the oven to 350°F [180°C]. Grease an 11 by 15½ in [28 by 39 cm] rimmed baking sheet with softened butter.

2. To make the cake, in a large mixing bowl, sift together the flour, granulated sugar, cinnamon, and baking soda.

3. In a small saucepan over medium heat, melt the butter and lard. Sift in the cocoa powder and whisk to combine. Stir in 1 cup [240 ml] of water and bring to a boil, then quickly remove from the heat and pour over the flour mixture. Stir to combine.

4. In a separate medium bowl, beat the eggs gently and mix in the buttermilk and vanilla. Pour the egg mixture into the flour mixture and fold it gently, being careful not to overmix. Transfer the batter to the greased baking sheet and bake on the center rack of the oven for 15 to 20 minutes until firm in the center but not crisp.

5. To make the icing, in a small saucepan over medium heat, melt the butter with the milk. Whisk in the cocoa powder. Bring to a boil and slowly add the confectioners' sugar, whisking to incorporate. Remove the icing from the heat, then whisk in the vanilla. Pour the icing over the cake while it's hot. Once it cools to room temperature, slice into squares and serve.

The cake can be stored out on the counter covered with plastic wrap or foil for 4 days or in the refrigerator for up to a week. Our family has also cut it into brownie size squares and placed it in the freezer. (This was initially an attempt at not eating it all at once, but then we discovered that it also tastes amazing frozen!)

SUMMER

Summer Dinner Parties with Our Favorite Chefs and Humans

The majority of the cheese we make goes to the farmers' markets—that way, it ends up in the eater's hand as soon as possible, we are sure to make a bit of a profit, and we get to hang out with our farmer friends and incredible cheese-loving supporters. Beyond the markets, much of our cheese goes to Bay Area restaurants and small local cheese shops. We are blessed to live in an area where we don't have to do any advertising, as restaurant owners and staff—including the chefs—often shop at farmers' markets. They seek out what tastes amazing; then they visit the farm to ensure that we treat the animals well and are taking care of our pastures.

We are so humbled and proud when we eat at one of our favorite restaurants and find our cheese not only on the cheese board but also in a dish we would never have thought of. Jessica visited the restaurants of some of our favorite chefs; they graciously taught her their recipes, and now we are passing them along to you. Enjoy!

When the stars align, the chefs bring their teams to the farm for a night to be remembered. The chefs we know followed that profession out of their love for delicious, locally grown ingredients that they creatively turn into something that evokes pleasure and a memory that you will hold on to. David and I enjoy nothing more than sitting outside at a long table and eating with old friends and new. Nan McEvoy of McEvoy Ranch in Petaluma used to have epic harvest feasts—there was so much joy throughout the day. Our neighboring farm, True Grass, hosts a community potluck once a month, and it is the highlight of the week for so many. We had the same experience in Italy at a table that went on for a block.

We believe that you don't have to own a farm to replicate this feeling of a pleasurable feast with friends, neighbors, or someone you barely know. The goal is to simply celebrate the seasons, the farmers, the makers, and the humans around you that make life so wonderful. This can happen with a simple picnic in a park, in a backyard with a string of lights, or at your own dining room table with candles and flowers. Take a walk around your yard or neighborhood to find flowers or branches for your tablescape (ask permission from neighbors first). Keep it simple and throw in a bit of whimsy (your grandmother's china, linens from a flea market, cozy blankets, flowers that you dried). The main goal is to have fun, dig into your creative side (we all have it somewhere), make food you love that you can source close by, and then invite loved ones. Putting our phones down, eating good food, engaging in wonderful conversation under the night sky—it just doesn't get any better than this.

TO START
TO START
TO START
TO START
TO START

TO START
TO START
TO START
TO START
TO START

The Perfect Summer Cheese Board

There is one rule for making a cheese board: Have fun! On the farm we make several cheese boards a week for folks who visit the farm. When building a Tomales Farmstead Creamery cheese board, we select seasonal cheeses while gathering other ingredients such as the edible flowers, herbs, and fruit. We are fortunate to have apples, peaches, plums, and blackberries growing on the farm. If we don't have fruit in season, we pick up fruits from the farmers' market or local farm stand. In the winter we grab a jam or marmalade that we made and have stored for the winter. We use fresh-out-of-the-oven baguettes from our town bakery, Route One Bakery and Kitchen. The soft bread is excellent for cheeses served with accompanying spreads. When including crisp crackers, standing them upright in a glass on the board creates height and interest.

Tamara recommends keeping color and texture in mind when selecting your accompanying fruit, flowers, herbs, and other garnishes. There is nothing appetizing about a beige board! If you don't have access to edible flowers, consider adding a variety of herbs in large bunches between the cheeses, or placing the cheese over leaves from your garden. Make sure to use a separate knife for each cheese and spread to prevent flavors from mingling. Lastly, as you present the board to your guests, share a bit about each cheese and where it comes from.

MAKES 6 TO 8 SERVINGS

3 or 4 types of cheese, 2 oz [55 g] per cheese (a mix of soft, semisoft, crumbly, and thinly sliced hard cheeses), at room temperature (fresh cheese can stay refrigerated)

1 fresh baguette, sliced diagonally about ¼ in [6 mm] thick

Tall crackers or thin breadsticks

Fruit, such as persimmon, pear, or apple, thinly sliced

3 oz [85 g] dried fruit, such as figs or apricots, or fresh seasonal berries

2 oz [55 g] accompanying spread, such as honey, chutney, or jam

2 large fresh rosemary sprigs, for garnish

Edible flowers, such as nasturtiums, pansies, borage, lavender, chive blossoms, or rose petals, for garnish

1. On a cheese board or serving platter, arrange the selected cheeses, ensuring they are widely spaced on the board. Arrange the baguette slices so they fan in strips around the cheeses. You may want to serve additional baguette slices in a cloth-lined basket on the side of the board. Arrange tall crackers standing upright in a glass. Fill in the gaps with a mixture of fresh sliced and dried fruit. Spoon the spread into a small bowl and place it on the board. Lastly, nestle fresh herb and edible flowers around the cheeses and borders of the board. Serve immediately.

Roasted Sungold Tomato and Teleeka Bites

This dish couldn't be simpler: Tomatoes are roasted with herbs, olive oil, and salt, spooned atop cheesy bread, and garnished. This is a beautiful starter and pairs well with a crisp rosé on a warm summer evening. At the height of summer, we recommend reserving a handful of fresh tomatoes to garnish the roasted ones, adding complexity to the flavors and texture of the bites.

Sungolds are a low-acidity and slightly sweet variety of cherry tomato that are ripe at the height of summer when they reach a pale apricot color. When they're in season, we buy them compulsively. They're also quite easy to grow, reaching up to 10 ft [3 m] in height and ripening in about three months from green to gold.

MAKES 6 TO 8 SERVINGS

2 cups [about 285 g] Sungold cherry tomatoes or other sweet cherry tomatoes, 6 to 8 reserved for garnish

2 Tbsp extra-virgin olive oil

1 bunch fresh basil, tied with twine, plus 4 to 6 fresh basil leaves, julienned, for garnish

1 fresh baguette or country bread loaf, sliced and warmed

4 oz [115 g] triple cream cheese (like our Teleeka) or burrata cheese

¼ tsp Maldon sea salt

1. Preheat the oven to 375°F [190°C].

2. In a small baking dish, add the tomatoes along with the olive oil and the bunch of basil. Bake for 30 minutes, stirring halfway through, until the tomatoes burst and create a thick sauce. Remove from the oven and carefully discard the bunch of basil.

3. Arrange the warm baguette slices on a serving platter and arrange slices of the cheese on each one, gently spreading the cheese with a knife. Spoon 1 Tbsp or so of the tomato mixture over each slice of bread. Slice the fresh tomatoes in half and top the bread with them. Sprinkle the platter evenly with the salt and julienned fresh basil to serve. The bites can remain at room temperature for up to 3 hours and are best eaten the day they are made.

Whipped Garden Herb and Liwa Mousse

This mousse is my favorite pairing for gorgeous fresh vegetables at a dinner party or out at a picnic. When we packed this for a sunset picnic with friends on the Bolinas Fairfax Ridge, it was devoured within moments. No one could quite tell what it was—too light to be a dip, too herbal and fresh to be a cheese spread, but too creamy to be dressing. And the mystery lived on until now. Enjoy this garden herb mousse, and see if your guests can guess what's in it.

MAKES 6 SERVINGS

½ ripe Hass avocado

5 oz [140 g] fresh goat cheese (like our Liwa)

Zest and juice of 1 large Meyer lemon, plus more juice as needed

1 whole peeled garlic clove

1 Tbsp extra-virgin olive oil, plus more as needed

1 cup [40 g] fresh chopped parsley

1 cup [40 g] fresh chopped mint leaves

½ tsp kosher salt

¼ tsp freshly ground black pepper

Mixed vegetable crudités, for serving

Fresh bread, for serving

1. In a blender, add the avocado, cheese, lemon zest and juice, garlic, olive oil, parsley, mint, salt, and pepper and blitz on high speed until combined evenly into a light green whipped mousse, about 3 minutes.

2. Add additional olive oil and lemon juice, a few drops at a time, if more moisture is needed. Taste for seasoning. Serve with crudités and fresh bread, or as a spread on sandwiches. The mousse can be stored in an airtight container in the refrigerator for up to 6 days.

Kale Salad with Cashews and Koto'la

This kale salad instantly becomes a meal when served over a bed of steamed brown rice or orzo. The flavors are punchy and addictive, and the raw kale is incredibly good for you! Wash your hands and then spend a good 5 minutes or more gently massaging the dressing into the kale. The acid from the aminos and lemon will tenderize the leaves and make them incredibly soft and a pleasure to consume.

MAKES 4 SERVINGS

1 large bunch green curly kale, tough stems removed, leaves torn into bite-size pieces (about 6 cups [90 g])

2 tsp Bragg Liquid Aminos

2 tsp lemon juice

1 tsp sesame oil

1 tsp tahini paste

¼ cup [60 ml] extra-virgin olive oil

¼ cup [35 g] cashews, roughly chopped and lightly toasted

3 oz [85 g] feta-style cheese (like our Koto'la)

2 oz [55 g] pomegranate seeds

1. Add the kale pieces to a large serving bowl.

2. In a small bowl, combine the liquid aminos, lemon juice, sesame oil, and tahini. Slowly whisk in the olive oil until lightly emulsified. Pour the dressing over the kale and massage it into the leaves for 3 to 5 minutes, until the toughness of the leaves is nearly gone and they feel tender.

3. Add the cashews to the bowl, then crumble in the cheese. Top with the pomegranate seeds and serve. We recommend eating this dish the same day it's made—it can be kept in the refrigerator for up to 4 hours until ready to serve.

Charred Watermelon Salad with Basil and Koto'la

Koto'la is a brined feta-style goat cheese, crumbly and creamy, with a salty bite that when paired with fresh fruit creates a brightness perfect for a summer side. Whip up this salad when the grill is already hot and your menu is in need of color and freshness. We recommend reserving a few whole basil leaves to garnish along with some Koto'la to crumble on each serving.

1. Preheat the grill to 400°F [200°C].

2. Lightly brush each watermelon wedge with olive oil and place on the grill for 2 to 3 minutes per side, until char marks appear. Arrange on a large platter.

3. Crumble 3 oz [85 g] of the cheese into a small blender. Add the sour cream, basil, mint, lemon juice, salt, and pepper. Blend until creamy and uniform in color, about 1 minute.

4. Lightly drizzle the mixture over the watermelon. Tear the basil leaves across the platter and top with some freshly ground black pepper and the remaining 2 oz [55 g] of cheese. Serve immediately. We recommend eating this dish the same day it's made.

MAKES 8 TO 10 SERVINGS

1 watermelon, quartered, then cut into 1 in [2.5 cm] triangles

Extra-virgin olive oil

5 oz [140 g] feta-style cheese (like our Koto'la)

¼ cup [60 g] sour cream

1 heaping cup [12 g] basil leaves, plus more for garnish

1 heaping cup [12 g] mint leaves

Juice of ½ lemon

½ tsp fine salt

¼ tsp freshly ground black pepper, plus more to season

Stone Fruit and Flowers

This salad looks like a painting and is bursting with flavor. Fresh summer stone fruit is the main subject here. In place of lettuces are copious amounts of fresh herbs and edible flowers. And for the right balance, there is plenty of toasted bread, creamy cheese, and salty prosciutto. You can make your own croutons easily by dicing crusty sourdough bread into 1 in [2.5 cm] cubes, tossing in olive oil and spices, and toasting the cubes in the oven to your preferred crispness. If you don't already grow edible flowers, this salad will inspire you to add them to your garden.

1. To make the dressing, in a salad bowl, whisk together the olive oil, vinegar, mustard, salt, and pepper and set aside.

2. To assemble the salad, add the nectarines, apricots, and croutons to the dressing and toss. Crumble in the cheese, then add the prosciutto, half of the flowers, half of the parsley, and half of the mint and toss until evenly coated. Top with the remaining herbs and flowers to serve. This salad can be stored without the dressing in an airtight container in the refrigerator for up to 4 hours.

MAKES 6 SERVINGS

DRESSING

½ cup [120 ml] extra-virgin olive oil

3 Tbsp sherry vinegar

1 Tbsp Dijon mustard

½ tsp Maldon sea salt

¼ tsp freshly ground black pepper

SALAD

2 ripe nectarines, pitted and sliced into thin wedges

2 ripe apricots, pitted and sliced into thin wedges

2 cups [80 g] sourdough croutons

2 oz [55 g] triple cream cheese (like our Teleeka)

4 oz [115 g] thinly sliced prosciutto

1 cup [12 g] edible flowers, such as borage, pansies, garlic chives, and nasturtiums

¼ cup [10 g] roughly chopped fresh parsley

3 Tbsp roughly chopped fresh mint

3 Tbsp torn fresh basil leaves

Pansy

Smoked Trout on Crackers with Liwa and Olives

We all have that snack that stands in as a quick meal, usually eaten over the kitchen sink or on a napkin while attempting to multitask with chores, children, or other daily labors. It's barely a recipe, but it's so good we had to include it. This dish can be dressed up for company as well, served on toasted slices of baguette, or portioned as a tartine sandwich for lunch on a slice of sourdough bread with a mixed green salad.

1. Spread the cheese in a thin layer across the crackers. Sprinkle the chopped olives evenly over the top.

2. With a fork, gently flake the trout into thin pieces. Spread the trout atop the goat cheese and olives. Serve immediately. We recommend eating this dish the same day it's made.

MAKES 2 SERVINGS

2 Tbsp [55 g] fresh goat cheese (like our Liwa)

2 large sheet crackers, such as matzo or La Panzanella (about 4 oz [115 g])

¼ cup [60 ml] cold Castelvetrano olives, pitted and roughly chopped

6 oz [170 g] smoked trout packed in oil, rinsed and patted dry

The Cheesemaker's Proja [Proi-ya]

This recipe is contributed by Nick Campbell, Tomales Farmstead Creamery Head Cheesemaker.

Each time my mother took the tub of cornmeal off the pantry shelf, I eagerly waited for her to share her memories of the home-grown corn she enjoyed as a child in the former Yugoslavia (now Montenegro). In autumn, her family brought dried corn in from the fields to be shucked and shelled around the warmth of the woodstove. Then her aunt would fill a sack with these golden drops of Montenegrin sunshine, tie them to a saddle, and ride her horse to the miller to turn them into fine cornmeal. The true magic of the fresh cornmeal was revealed when turned into proja—a crusty loaf of bread, dotted with bits of the homemade cheese that her family made daily in the warm months that was stored to feed themselves in winter. This cornbread often made up the entire meal, to be eaten with a cold glass of mouth-puckeringly sour yogurt. With wheat bread now more available, some consider this traditional cornbread a symbol of poverty. But its incredible flavor—and the nostalgia it invokes—mean proja endures as a modern staple.

This cornbread is unlike the sweet corn cake offered with chili or at various American summer holiday meals. It is intensely savory and crisp, best eaten hot. If at all possible, try to get stone-ground cornmeal for this recipe. The flavor of freshly ground cornmeal is so incredibly different from its store-bought counterpart that they seem to be two completely different foods. If you can't find cornmeal freshly ground, adding a little extra cheese can make up for it. The secret to the crisp crust on this cornbread is a cast-iron skillet and a fair amount of fat preheated in the oven so the batter instantly fries as it hits the hot pan.

MAKES 8 SERVINGS

¼ cup [55 g] lard, bacon grease, or ghee

2 cups [280 g] stone-ground yellow cornmeal

1 Tbsp baking powder

2 tsp kosher salt (this can be adjusted depending on how salty the cheese is)

2 large eggs, beaten

1½ cups [360 ml] buttermilk, kefir, or clabbered raw milk (see Note)

8 oz [230 g] brined feta-style cheese (like our Koto'la or El Greco), crumbled

1. In a 10 in [25 cm] cast-iron skillet, add the lard, then place the skillet on the middle rack of the oven. Preheat the oven to 475°F [250°C]. It is best to make the cornbread batter when the oven is fully preheated so the baking powder fully leavens the bread as it bakes.

2. In a large bowl, whisk together the cornmeal, baking powder, and salt. Create a well in the center of the mixture and add the eggs and buttermilk. Beat well until the batter is free of lumps. Fold in the crumbled cheese right before baking the cornbread so that it stays evenly distributed throughout the batter.

3. Remove the preheated skillet from the oven and carefully pour in the batter, scraping the bowl with a spatula. The batter should immediately begin to sizzle in the heated fat. Smooth the top of the batter and return the skillet to the oven.

4. Bake for 30 minutes, rotating the skillet halfway through, until the cornbread is deeply golden-brown and the sides of the bread have pulled away from the skillet. Remove from the oven and serve while warm. It can be stored in an airtight container at room temperature for up to 3 days.

NOTE

Despite the health benefits of raw milk, raw milk has not been pasteurized and can pose the risk of illness. Please purchase your milk from trusted sources and follow safe practices.

MAINS
MAINS
MAINS
MAINS
MAINS

MAINS
MAINS
MAINS
MAINS
MAINS

Toluma Ploughman's Lunch

The traditional English ploughman's lunch was popular with nineteenth-century farmers in the English countryside. It became even more popular across pub menus when cheesemakers started using it in their advertising in the 1950s. The ploughman's lunch is a deconstructed platter of bread, cheese, eggs, and vegetables; think of this as the reconstructed version, in which jammy eggs spill over caramelized veggies in a lemony tahini sauce. Serve this with a crusty baguette and a cold pint of beer, just like they do in the pubs.

1. To make the lunch bowl, bring a medium saucepan of water to a rolling boil. Add the potatoes and boil for approximately 20 minutes.

2. While the potatoes are boiling, make the tahini dressing in a large salad bowl by whisking together the olive oil, lemon juice, lemon zest, sherry vinegar, tahini, garlic, salt, and pepper. Set aside.

3. After 20 minutes, add the romanesco and carrots to the pot and blanch with the potatoes for approximately 5 minutes. Drain all of the vegetables and set aside.

4. Refill the pot and bring the water to a boil. Fill a medium bowl with ice water. Add the eggs to the boiling water and lower the heat to a gentle simmer. Cover and let cook for 6 minutes. Transfer the eggs to the bowl of ice water.

5. Heat a large skillet over medium heat and warm the olive oil until it shimmers. Add the cooked potatoes, romanesco, and carrots along with a pinch of salt and toss to coat. Sauté until the vegetables are slightly caramelized, about 5 minutes. Add the cheese, lemon zest, and lemon juice. Remove from the heat.

6. To serve, toss the vegetables in the tahini dressing and plate in shallow bowls. Peel the eggs gently under running water, pat dry, and slice 2 eggs in half over the top of each bowl. Garnish with the parsley, mint, a pinch of salt on each yolk, and a light grind of pepper. Serve immediately with a few slices of baguette. We recommend eating this dish the same day it's made.

MAKES 4 SERVINGS

LUNCH BOWL

1 cup [125 g] sliced fingerling potatoes, sliced into ½ in [13 mm] chunks

2 cups [200 g] romanesco florets

1 cup [130 g] chopped carrots, cut into 2 in [5 cm] chunks on the diagonal

8 large eggs

2 Tbsp extra-virgin olive oil

Kosher salt

2 oz [55 g] feta-style cheese (like our Koto'la)

Zest and juice of 1 lemon

2 Tbsp roughly chopped fresh parsley

1 tsp roughly chopped fresh mint

Freshly ground black pepper

TAHINI DRESSING

¼ cup [60 ml] extra-virgin olive oil

1 Tbsp lemon juice

1 tsp lemon zest

½ Tbsp sherry vinegar or white wine vinegar

2 tsp tahini

1 garlic clove, smashed

½ tsp kosher salt

¼ tsp freshly ground black pepper

Baguette, sliced, for serving

Dijon Grilled Cheese

There is nothing quite like the golden crunch, creaminess, and bite of a grilled cheese sandwich. The Dijon puts the experience of this dish somewhere between a Cubano and croque monsieur, but I must call it a grilled cheese sandwich because it is quintessentially Californian when served on fresh sourdough bread with creamy Teleeka cheese. Prosciutto and rosemary ensure this won't be mistaken for a kid's lunch—it is an elevated sandwich that can be served cut into 1 in [2.5 cm] bites with cocktails, or served with a green salad as a meal.

MAKES 1 SANDWICH

2 Tbsp salted butter, plus more for buttering

2 Tbsp Dijon mustard

2 thick slices sourdough bread

1 thin slice prosciutto

2 to 3 oz [55 to 85 g] triple cream cheese (like our Teleeka)

1 tsp roughly chopped fresh rosemary

1. In a cast-iron skillet over medium-high heat, lightly brown the butter. Meanwhile, spread a generous layer of Dijon mustard on one side of each slice of bread, and a layer of butter on the other side.

2. Add the bottom slice of bread to the pan, mustard-side up, and layer on the prosciutto. When the bread begins to crisp, layer on the cheese and top with the remaining slice of bread, mustard-side down. Gently flip the sandwich with a spatula. Lower the heat to low and cook until the cheese is softened.

3. Remove from the pan and slice diagonally before serving. Garnish with chopped rosemary. We recommend enjoying this dish the same day it's made.

Heirloom Tomato and Nasturtium Galette

In the summer, there is a natural impulse to ensure all the colors of the rainbow are represented in your cooking. Because we eat with our eyes first, food should be beautiful, and during tomato season, that is easy to achieve. Spend a little extra time picking out the most beautiful tomatoes you can find. The combination of fresh and cooked tomatoes here allows you to enjoy a surprise combination of complex flavors and textures without having to toil with a complex recipe. Dot the galette with beautiful nasturtiums, especially the whimsical round leaves, and you'll think this dish is almost too pretty to eat.

MAKES 8 SERVINGS

DOUGH

2½ cups [350 g] all-purpose flour, plus more for dusting

1 tsp kosher salt

1¼ cups [280 g] chilled unsalted butter, cut into ½ in [13 mm] pieces

¼ cup [60 ml] ice water, plus more as needed

FILLING

1½ lb [680 g] heirloom tomatoes of varying color, sliced ¼ in [6 mm] thick

½ cup [75 g] cherry tomatoes, halved

2 Tbsp sherry vinegar

½ tsp kosher salt

¼ tsp freshly ground black pepper

1 lb [455 g] fresh goat cheese (like our Liwa)

3 Tbsp honey

1 tsp extra-virgin olive oil

2 tsp grated lemon zest

3 Tbsp julienned fresh basil

½ cup [6 g] nasturtium flowers and leaves

1. Preheat the oven to 400°F [200°C] and line a baking sheet with parchment paper.

2. To make the dough, in a food processor, add the flour, salt, and butter and pulse to form pea-size crumbs. With the motor running, add the ice water just until the dough comes together. Transfer the dough to a lightly floured work surface. Shape into a 1 in [2.5 cm] thick disk, wrap it in plastic wrap, and refrigerate for 30 minutes to 1 hour.

3. To make the filling, in a large bowl, gently toss the heirloom and cherry tomatoes in the sherry vinegar, salt, and pepper, and set aside.

4. With a rolling pin, roll out the dough to form a 12 to 13 in [30.5 to 33 cm] circle and place it on the prepared baking sheet. In a small bowl, mix the cheese, honey, olive oil, and lemon zest. Spread the mixture evenly over the dough, leaving a 2 in [5 cm] border around the edges.

5. Arrange two-thirds of the tomatoes on top of the cheese mixture, alternating colors and sizes. Reserve the remaining tomatoes. Fold the edges of the dough over the filling, overlapping each fold slightly and covering the edges of the tomatoes. Bake for 45 minutes until the crust is golden brown, the cheese is melted, and there are some light char marks on the tomatoes. Add the remaining fresh tomatoes to the galette and top with the basil, nasturtium flowers, and nasturtium leaves.

6. Allow to cool slightly and slice into small wedges to serve. The galette can be stored in an airtight container in the refrigerator overnight. We recommend eating this dish the same day it's made.

Balsamic Salad Pizza

My first job at the age of fifteen was bussing dishes at Small Shed, a little neighborhood restaurant in southern Marin. I had never seen a pizza piled high with salad until I worked there, and once I tried it, it was my go-to staff meal. It's a great big salad you can fold up in a thin, cheesy flatbread. The balsamic drizzle tossed onto the salad adds a sweetness that complements the salty cheese and fresh lettuces. It's so light and delicious that it's barely fair to call it a pizza.

MAKES 4 SERVINGS

Semolina flour, for dusting

1 cup [240 ml] balsamic vinegar

¼ cup [50 g] brown sugar

½ recipe Sourdough Pizza Crust (page 114)

3 cups [60 g] mixed baby spring lettuces

¼ cup [60 ml] extra-virgin olive oil

4 oz [115 g] creamy sheep milk cheese (like our Bossy)

1. Preheat the oven to 550°F [290°C] for 1 hour with the rack positioned in the top third of the oven. Line a baking sheet with parchment paper, sprinkle with semolina, and set aside. If you have a pizza stone, preheat the stone in the oven.

2. In a small saucepan over medium heat, mix together the vinegar and brown sugar, stirring constantly until the sugar has dissolved. Bring to a boil, then lower the heat to a low simmer, stirring occasionally, until the glaze is reduced by half and coats the back of a spoon, about 20 minutes. Let the glaze cool, then pour into a lidded jar. Seal and store in the refrigerator until ready for use (remaining glaze can be stored in the refrigerator for up to 10 days).

3. With a rolling pin, roll out the pizza dough until it's about ⅛ in [4 mm] thick. Place the dough on the prepared baking sheet. Spread the cheese into an even layer on the dough. Place the baking sheet in the oven (or transfer the dough to the pizza stone) and turn the oven to broil for the first 4 minutes, and then back to bake for 15 minutes, until the crust is golden, the edges are crisp, and the cheese is melted.

4. While the pizza is in the oven, dress the lettuce. In a large bowl, toss the lettuce with the olive oil and just enough (2 to 3 Tbsp) balsamic glaze to coat it.

5. Let the pizza cool slightly on the baking sheet or a cutting board. To serve, slice into wedges and place on a platter. Top with the cool salad and serve immediately. We recommend enjoying this dish the same day it's made.

SOURDOUGH PIZZA CRUST

There is a wonderful outdoor pizza oven at the farm that we often use for pizzas. Since not everyone has a pizza oven, all the pizza recipes in this book are designed to be made with indoor ovens. This is a basic recipe for pizza dough you can make at home. We recommend measuring all ingredients by their weight in grams using a small home kitchen scale. This will ensure greater precision depending on the type of flour you use.

MAKES 2 PIZZA CRUSTS

288 g [2 cups] Type 00 white flour or all-purpose flour

32 g [3 Tbsp] whole wheat flour

215 ml [⅞ cup] ml water

2 g [½ tsp] diastatic malt (optional)

6 g [1 tsp] salt

1 Tbsp extra-virgin olive oil, for greasing the bowl

48 g [¼ cup] sourdough starter, ripe/discard

1. Make the dough 1 day before you plan to bake. In a large mixing bowl or stand mixer, add the 00 flour, whole wheat flour, water, and malt, if using. Mix on medium speed until the dough is smooth but still elastic, about 4 to 5 minutes. Transfer the dough to a bulk fermentation container such as a large tub or mixing bowl, cover, and leave in a warm place. Every 30 minutes for 2½ hours, reach into the fermentation container and pick up and pull a corner of the dough into the middle to stretch and fold the dough. Bubbles should begin forming across the surface.

2. Remove the dough from the container and lightly oil the interior with olive oil. Grasping from the bottom, tighten and roll the dough on the counter into a ball and transfer it back to the oiled container, seam-side down. Cover the dough directly with plastic wrap to prevent a crust from forming, then cover the container with a damp towel and transfer it to the refrigerator overnight.

3. The next day, divide the dough into two 10¼ oz [290 g] pieces. If you plan to make only one pizza, you can tightly wrap and freeze the second ball of dough for up to 3 months.

4. Shape the dough into a very tight ball with a closed seam on the bottom. To do this, hold the dough in one hand and pinch the sides into the center with the other. Rotate and repeat until all sides have been tightly pinched to the center, creating a closed seam; the underside should be tight and smooth. Transfer to a pizza dough tray or baking sheet and cover with a large bowl. Proof the dough on the counter at around 75°F [24°C] for 5 to 6 hours. When fully proofed, the dough will have relaxed outward and be soft to the touch. If using the dough soon, preheat your oven. Alternatively, you can place the dough back into an oiled, airtight container in the refrigerator until the next day or freeze the dough until ready for use.

5. To bake the pizza, follow the instructions from the pizza recipe of your choice.

Meyer Lemon and Honey Pizza

Double 8 Dairy is a close neighbor of Toluma Farms, raising water buffalo just down the road in Petaluma. They are one of only two water buffalo dairies in California, producing milk, cheese, and gelato. In this recipe, the combination of their Fior di Latte cheese with our Atika is perfectly paired with caramelized lemons and earthy honey. If you can't get your hands on their cheese, burrata will do.

MAKES 4 SERVINGS

Semolina flour, for dusting

½ recipe Sourdough Pizza Crust (page 114)

4 oz [115 g] burrata-style cheese (like Double 8 Dairy's Fior di Latte)

1 Meyer lemon, very thinly sliced

2 Tbsp wildflower honey

2 oz [55 g] Parmesan-style cheese (like our Atika) or Pecorino Romano

3 Tbsp fresh torn basil

1. Preheat the oven to 550°F [290°C] for 1 hour with the rack positioned in the top third of the oven. Line a baking sheet with parchment paper, sprinkle with semolina, and set aside. If you have a pizza stone, preheat the stone.

2. With a rolling pin, roll out the pizza dough until it's about ⅛ in [4 mm] thick. Place the dough on the prepared baking sheet. Spread out the burrata-style cheese into an even layer on the dough, then arrange the lemon slices on top. Place the baking sheet in the oven (or transfer the dough to the pizza stone) and turn the oven to broil for the first 4 minutes, and then back to bake for 15 minutes, until the crust is golden, the edges are crisp, and the cheese is melted.

3. Let the pizza cool slightly on the baking sheet. Drizzle the honey, sprinkle the Parmesan-style cheese to melt, and when slightly cooled, sprinkle the fresh basil over the top. To serve, slice the pizza into wedges and place on a platter. This pizza can be stored in an airtight container in the refrigerator for up to 1 day (leftovers are good served cold, too!).

Peach and Serrano Ham Pizza

Peaches and prosciutto are a classic combination. And at the height of summer, there is nothing better than enjoying your stone fruit for breakfast, lunch, and dinner. This pizza is a savory treat with an herbal hit that showcases beautiful peaches at the peak of the season.

MAKES 4 SERVINGS

Semolina flour, for dusting

½ recipe Sourdough Pizza Crust (page 114)

4 oz [115 g] triple cream cheese (like our Teleeka)

1 ripe peach, pitted and sliced very thin

3 oz [85 g] Serrano ham or prosciutto, sliced into thin ribbons

2 oz [55 g] Parmesan-style cheese (like our Atika) or Pecorino Romano

3 Tbsp fresh parsley leaves

1. Preheat the oven to 550°F [290°C] for 1 hour with the rack positioned in the top third of the oven. Line a baking sheet with parchment paper, sprinkle with semolina, and set aside. If you have a pizza stone, preheat the stone.

2. With a rolling pin, roll out the pizza dough until it's about ⅛ in [4 mm] thick. Place the dough on the prepared baking sheet. Spread out the triple cream cheese into an even layer on the dough, then arrange the peach slices on top. Place the pizza in the oven (or transfer the dough to the pizza stone) and turn the oven to broil for the first 4 minutes, and then back to bake for 15 minutes, until the crust is golden, the edges are crisp, and the cheese is melted.

3. Let the pizza cool slightly on the baking sheet or a cutting board. Scatter the ham across the pizza in thin slices and sprinkle the Parmesan-style cheese to melt. When slightly cooled, sprinkle the parsley over the top. To serve, slice the pizza into wedges and place on a platter. This pizza can be stored in an airtight container in the refrigerator for up to 1 day (leftovers are good served cold, too!).

Lemon Bucatini with Atika

This recipe is contributed by Nico Van de Bovenkamp.

Tamara's son-in-law, Nico Van de Bovenkamp, loves nothing more than cooking for others; it is the way he finds joy and relaxes, and his family and friends benefit greatly from his love of cooking. This light and flavorful recipe of his is a fan favorite and is often served alongside pan-seared shrimp or scallops.

MAKES 6 SERVINGS

1 Meyer lemon

Kosher salt

12 oz [340 g] bucatini (or other long pasta)

½ cup [120 ml] heavy cream

5 Tbsp [75 g] unsalted butter

¾ cup [25 g] Parmesan-style cheese (like our Atika), finely grated

8 oz [230 g] feta-style cheese (like our Koto'la)

Freshly ground black pepper

1. Use a vegetable peeler to remove one-third of the lemon zest; slice into thin strips and set aside. Finely grate the remaining zest into a large pot. In a small bowl, squeeze about 2 Tbsp of lemon juice and set aside.

2. In a separate large pot, boil water, salt it heavily, and cook the pasta until molto al dente (slightly undercooked). Reserve about 1 cup [240 ml] of the pasta water before draining.

3. While the pasta cooks, add the heavy cream to the pot with the lemon zest and bring to a simmer over medium heat, stirring often. Lower the heat to low and whisk in the butter 1 Tbsp at a time until creamy. Add the pasta to the sauce along with ¾ cup [180 ml] of the pasta water.

4. Increase the heat to medium and slowly add the Parmesan-style cheese until the sauce is thick and creamy, tossing often until the pasta is cooked, about 3 minutes. (If the sauce looks too thick, add more pasta water a splash at a time.)

5. Finish by stirring in the reserved lemon juice and season with salt. Serve immediately and top with the reserved lemon zest strips, feta-style cheese, and pepper. We recommend enjoying this dish the same day it's made.

Teleeka and Summer Squash Lasagna

This recipe is contributed by Daniella Banchero, executive chef at Piccino.

There are a few consistent problems with lasagna: When it's good, everyone fights over the edge pieces, and when it's not good, no one gets terribly excited about sloppy piles of ground beef, red sauce, and rubbery slabs of mozzarella. Daniella Banchero, the executive chef at Piccino in San Francisco, has miraculously solved both problems in a single recipe, and it is the most delicious lasagna I have ever eaten.

It's unbelievably light on the inside and delightfully crispy on the outside. The sweet and slightly tart Sungold tomatoes cut through the creamy richness of the cheese, and the entire dish feels like you're eating a boldly flavorful pasta dish in the texture of a French patisserie dessert. Every bite is a perfect bite because rather than scooping from the baking dish to the plate, each slice is seared on all sides, making each serving a coveted edge piece! You will need to exercise restraint from devouring the entire pan.

MAKES 12 SERVINGS

PASTA DOUGH

2 cups [240 g] 00 Caputo flour

1 tsp kosher salt

2 large eggs

3 egg yolks

¼ cup [30 g] semolina flour, for dusting

continued

1. To make the pasta dough, on a clean and dry work surface, place the flour into a mound. Make a well in the center of the flour and add the salt, eggs, and yolks. Using a dinner fork, whisk the eggs, gently working outward to begin incorporating the flour. Use clean hands to continue incorporating the eggs into the flour until all the moisture is absorbed and no dry flour patches remain. Knead the dough for 5 to 8 minutes until very smooth. Wrap the dough in plastic wrap and refrigerate for 1 hour.

2. Divide the dough into 8 equal pieces. Press one piece into a flat disk. Using a pasta roller, roll out the dough, gradually moving from the thickest to thinnest setting. Dust the dough with semolina flour and place it on a baking sheet. Cover and refrigerate until you're ready to assemble the lasagna. Repeat with the remaining dough pieces.

continued

3. To make the béchamel, in a 2 qt [1.9 L] pot over low heat, warm the butter until fully melted but not foaming. Sift the flour into the butter and whisk vigorously to combine until the mixture is completely smooth and no lumps remain, forming a roux. Cook for 30 seconds until lightly golden, then set aside to cool completely.

4. In a 2 qt [1.9 L] pot over medium-high heat, warm the milk until scalded but not boiling. Slowly add the milk to the roux and return the pot to low heat. Whisk constantly until the texture is slightly thickened and coats the back of a spoon, about 5 to 7 minutes. Add the garlic, salt, and pepper. Stir to combine, then remove from the heat. The sauce should be fairly thick at this stage and may continue to thicken slightly as it rests. Béchamel sauce can be made in advance and stored for 2 to 3 days in an airtight container in the refrigerator. If making in advance, place a piece of plastic wrap directly on the surface of the sauce to prevent it from forming a skin.

5. To make the filling, lay out the zucchini and squash slices on a baking sheet. Sprinkle with salt and let it sit for 10 to 15 minutes to release the moisture, then pat dry with a paper towel before assembling.

6. Preheat the oven to 375°F [190°C].

7. To assemble the lasagna, butter a 9 by 13 in [23 by 33 cm] baking dish that is at least 3½ [9 cm] deep. Cut the pasta to fit the inside of the dish so that no space is remaining around the sides. Fit two pasta sheets side-by-side for the bottom layer. Add an extremely thin layer of béchamel, similar to a layer of butter on toast. Add a layer of zucchini and squash slices, overlapping just slightly. Add the next layer of pasta, followed by a very thin layer of ricotta and triple cream cheese. The order should go as follows:

Pasta
Béchamel
Zucchini and yellow squash
Pasta
Ricotta and Teleeka
Pasta
Béchamel
Zucchini and yellow squash
Pasta
Ricotta and Teleeka

8. Press down firmly on each layer of pasta as you assemble to keep the layers tightly packed. Keeping the layers of béchamel and cheese very thin will help ensure a light texture after baking. Cover the pan with aluminum foil and bake for 90 minutes, rotating the pan halfway through. Remove the foil after 90 minutes, top with the Parmesan-style cheese, and bake for another 15 minutes to let the cheese melt. Let the lasagna cool completely. It can be refrigerated at this point, covered, for 3 to 5 days.

BÉCHAMEL

½ cup [110 g] unsalted butter

1 cup [140 g] all-purpose flour

4 cups [960 ml] milk

2 tsp finely chopped garlic

1 tsp kosher salt

½ tsp freshly ground black pepper

FILLING

1 lb [455 g] zucchini, thinly sliced lengthwise on a mandoline

1 lb [455 g] yellow squash, thinly sliced lengthwise on a mandoline

Kosher salt

Unsalted butter, for greasing the pan

4 cups [960 g] ricotta, homemade (see page 123) or store-bought

8 oz [230 g] triple cream cheese (like our Teleeka)

3½ oz [100 g] Parmesan-style cheese (like our Atika)

9. While the lasagna bakes and cools, prepare the Sungold tomato sauce. In a roasting pan, add the tomatoes, 5 or 6 sprigs of the basil, the salt, and olive oil. Bake for 20 minutes or until the tomatoes are bursting and the pan is filled with juice.

10. Remove the basil and transfer the tomatoes to a food mill placed over a saucepan. Mill the tomatoes to remove the seeds and skins. Alternatively, transfer the tomatoes to a fine-mesh sieve and press them through with a wooden spoon. Add the butter to the saucepan and cook over medium heat for 5 to 10 minutes until slightly reduced. Set aside to cool slightly.

11. Preheat the oven again to 375°F [190°C].

12. Slice the room-temperature lasagna into 3 in [7.5 cm] squares so that each portion is almost an exact cube. In a 9 in [23 cm] skillet over high heat, warm the vegetable oil. Sear all sides of the lasagna for 1 to 2 minutes per side, until browned and slightly crisp. Set the lasagna onto a baking sheet and return it to the oven for 10 minutes to heat all the way through.

13. To serve, pour 2 to 3 Tbsp of the tomato sauce on the center of each plate. Center a seared lasagna slice on the tomato sauce and drizzle a bit more sauce over the top. The lasagna can be tightly wrapped and stored in an airtight container in the refrigerator for up to 4 days.

SUNGOLD TOMATO SAUCE

6 cups [960 g] Sungold cherry tomatoes or other sweet cherry tomatoes

1 bunch of fresh basil

1 tsp kosher salt

2 Tbsp extra-virgin olive oil

2 Tbsp unsalted butter

2 Tbsp vegetable oil, for searing

HOMEMADE RICOTTA

1. In a large pot over medium heat, warm the milk and cream until the mixture starts to get foamy and steam. Remove from the heat before it begins to boil. A thermometer should read 200°F [95°C]. Off the heat, add the sugar, salt, lemon juice, and ascorbic acid. Stir gently to combine. Let the mixture rest for 15 minutes, in which time curds should be visibly forming and separating from the watery and slightly yellow whey.

2. Set a fine-mesh sieve lined with cheesecloth over a large bowl and pour the mixture through the sieve to strain the curds from the whey, discarding the whey. Let the curds drain for 45 minutes. Scoop the curds from the cheesecloth into a container to store. The ricotta cheese can be stored in an airtight container in the refrigerator for up to 1 week.

MAKES 3 TO 4 CUPS [720 TO 960 ML]

2 qt [2 L] whole milk

4 cups [960 ml] heavy cream

1 Tbsp granulated sugar

2 tsp kosher salt

1½ tsp freshly squeezed lemon juice

¾ tsp ascorbic acid

Truffle Roast Chicken

In the early 1900s, poularde en demi-deuil became a popular dish in France—a braised chicken studded with paper-thin slices of black truffles tucked under the skin of the bird. It literally translates to "chicken in half mourning" because it's only partially dressed in black. I love the idea of truffle chicken, but finding fresh truffles is not easy for mere mortals, and braising in liquid sounds like you're gearing up for a soggy-skinned bird.

This is our take on the old-fashioned dish, using truffle butter, which can now commonly be found in specialty food stores across the country. You roast the chicken on a bed of sliced fingerling potatoes, which fry in the juices of the bird; then reduce the pan drippings for a luxe version of truffle fries. There is no question that this is a rich and romantic dish, so consider serving it with a light salad and perhaps a nice bottle of Champagne.

MAKES 4 SERVINGS

1 lb [455 g] fingerling potatoes, cut into 1½ in [4 cm] pieces

4 lb [1.8 kg] whole organic roasting chicken, at room temperature

¼ cup [60 g] truffle butter, at room temperature, plus 2 Tbsp for the pan drippings

4 or 5 fresh thyme sprigs

4 or 5 fresh rosemary sprigs

¼ cup [60 g] salted butter, melted

2 Tbsp kosher salt

1. Preheat the oven to 425°F [220°C].

2. Place the potatoes in a large cast-iron baking dish or roasting pan.

3. Remove the neck and giblets if they've been left in the chicken cavity and pat the chicken completely dry with paper towels. Gently run your fingers under the skin of the chicken breasts and spread the truffle butter across the breasts, working it into a thin layer under the skin.

4. Stuff the cavity of the chicken with the thyme and rosemary, truss the legs of the chicken snugly (see page 70), and place the whole chicken on the bed of potatoes. Drizzle the salted butter over the top of the entire chicken, then sprinkle evenly with the salt.

5. Roast for about 90 minutes. The skin should be slightly blistered and crisp across the top of the chicken and the thick part of the thigh should read 165°F [75°C] on a meat thermometer. Place the chicken on a carving board and tent with aluminum foil.

6. Return the pan with potatoes to the stovetop and, over medium-high heat, reduce the pan drippings until you can see the bottom of the pan when you stir the potatoes, about 3 to 5 minutes. Remove from the heat and add the remaining 2 Tbsp of truffle butter to the pan, stirring to melt.

7. To serve, carve the chicken and place it on a serving platter. Pour the drippings and potatoes over the top of the chicken to serve. The chicken can be stored in an airtight container in the refrigerator for up to 4 days.

Cornish Game Hen Dijonnaise

When I was growing up, one of my grandpa's favorite dishes was chicken Dijon: two baked boneless, skinless chicken breasts smothered in a Dijon mustard sauce with dill. I believe, somewhat controversially, that there are more joyful ways to prepare chicken than boneless, skinless chicken breasts, so this is my take on his classic recipe.

I love to roast Cornish game hens when I can find them. They are tender, juicy, and unbelievably cute. I swapped out the dill for fennel (bulb and fronds), which adds a similar quality with a little extra sweetness. If you and your dinner guests have big appetites, it's fun to serve one hen per person. However, when carved and served over polenta or with side dishes, two hens can easily feed four people.

MAKES 4 SERVINGS

1 fennel bulb (about ½ lb [230 g]), roughly chopped, ¼ cup [5 g] chopped fronds reserved for garnish

2 Cornish game hens, patted dry and at room temperature

1 lemon, halved

4 fresh rosemary sprigs

5 Tbsp [75 g] Dijon mustard

3 Tbsp extra-virgin olive oil

2 tsp kosher salt

1 Tbsp unsalted butter

1 Tbsp freshly ground black pepper

1. Preheat the oven to 400°F [200°C].

2. In a 6 by 8 in [15 by 20 cm] baking dish or 9 in [23 cm] cast-iron skillet, spread the chopped fennel in a single layer. Place the hens on top of the fennel with some space between them so they aren't crowded. Stuff the cavity of each hen with half a lemon and 2 sprigs of rosemary. Mix 2 Tbsp of the Dijon with 2 Tbsp of the olive oil and brush the top of each hen evenly. Sprinkle the salt evenly over each hen and truss the legs with cotton string (see page 70).

3. Bake for 45 minutes, turning halfway through to ensure even browning. When removed from the oven, insert a knife into the space between the thigh and body. If the juices run clear, the hen is done. Turn the hens vertically when removing from the pan to release any additional juices before placing them on a carving board. Tent with aluminum foil to keep warm.

4. Transfer the juices, fennel, and any remaining contents from the baking dish into a large saucepan. Add the remaining 3 Tbsp of Dijon to the saucepan along with the remaining 1 Tbsp of olive oil, the butter, and pepper. Simmer gently over medium heat for 5 to 7 minutes until slightly reduced and thickened.

5. To serve, carve the hens and arrange on a serving platter or plate individually. Spoon the sauce and fennel over the hens and garnish with fennel fronds to serve. The hens and sauce can be stored in an airtight container in the refrigerator for up to 5 days.

An Alternative View of Cheese: CULTURAL CONTEXT MATTERS

NICK CAMPBELL, HEAD CHEESEMAKER

I was raised in a family that treated cheese as sustenance. Thick, snowy white slabs of imported Bulgarian sheep feta were laid out at nearly every meal; in the United States, it was the closest thing that my Serbian family from Montenegro could procure to invoke the brined white cheese of our mountain village. It was an alternative to meat; a flavoring and enrichment for our heavy, starchy cornmeal mush; and the quickest way to appease a whining child looking for a snack. At our summer meals, my grandmother's fresh-picked tomato, cucumber, and onion salad was made a muddy pink by shreds of feta mixed into the cooling, pungent juices released by the vegetables and dressing. I still shudder at one memory: All morning, my grandmother would labor over thin little pancakes for us, sprinkling sugar inside each one and rolling them up, only for some of my more adventurous cousins to unroll the pancakes and add in bits of salty white cheese as I looked on in horror at their strange exploration of sweet and savory.

Cheese was a beloved afterthought, a known staple that we always took for granted. Dairy was nothing novel to us; it just made sense that our immigrant family would keep consuming versions of products our ancestors spent innumerable hours making in order to survive the long Montenegrin Alpine winters. My mother and aunts still make yogurt at home because much of the American yogurt available is too bland for our tastes. There was a great joy in seeing the ritual that went into testing the heat of the milk with a clean little finger, the care with which the designated yogurt pot was wrapped in blankets brought from the homeland, not to be touched until the whey began to run from the yogurt and a beautiful lactic smell emerged when uncovered. The large metal tins of feta we kept in the garage gave the same sense of comfort and assurance: Milk in its many forms would always be there to feed us.

When I became more interested in food outside of my cultural bubble, I was startled by the way cheese is treated in gourmet circles. The reverence and cult following for Gruyère, the look of wonder and awe when the ooze flowed from an imported washed-rind cheese, even the theatrics of hiding a plate of pasta under a snowfall of grated Parmesan—none of these made very much sense to me. It took much time and lots of adventurous spirit to venture into this world of tiny blocks of expensive cheese, and I am still learning, knowing full well that I will never learn it all.

My ancestors are what drew me to make cheese. Although very little of what I do in my work is remotely similar to how my grandmother's family made cheese from their herds, I feel a distinct connection and satisfaction every time I cut a new batch of curd. The smell of whey reminds me of the stories I heard about the little wooden shack in the mountains where my great-grandmother made cheese and other dairy products daily in the spring and summer. She used a copper kettle and an open-hearth fire, dipping her finger to feel if the milk was at the right temperature. Today, I have a heated steel vat that can tell me the temperature of the milk within a tenth of a degree. She had to kill a calf every spring to make rennet; I now order mine online. The differences abound, but I still feel a deep connection with this woman I never knew. It might be wishful thinking, but I feel that she's passed something down her line to me, a predisposed notion of what milk and cheese mean to me.

There's a practicality in my point of view that I don't often see in this field; it must come from the cultural context I was raised in. I know my outlook can go against the grain, and I am interested and willing to delve into the way others look at artisan cheese. I make cheese primarily to feel the connection with my family roots and because I enjoy creating nourishment for others. Making food is an important calling. I am at a point in my thinking where I can appreciate the story and origin of each cheese put onto a board, and it can be entertaining to join the never-ending quest to try every piece of cultured milk in the display case. But nothing will ever impress me as much as a plate of feta set out among the other components of a meal, always there to sate and feed me.

Stemple Creek Rib-Eye with Rosemary Navy Beans

This recipe is contributed by Loren Poncia.

If you're going to enjoy a nice steak, make sure you support regenerative farming practices, like those used by our Tomales neighbors at Stemple Creek Ranch. Loren and Lisa Poncia use regenerative, organic agricultural practices, seeking to enhance and rehabilitate their entire ecosystem by focusing on soil health and increasing carbon in the land. Their pastured, free-range beef cattle and sheep are 100 percent grass fed and treated humanely for their lifetime on the ranch. You can truly taste the difference in this pan-seared rib-eye steak and can feel good about supporting small farmers.

Enjoy this beautiful steak dinner atop creamy navy beans and pair with a pinot noir or cabernet.

MAKES 2 SERVINGS

STEAK

¼ cup [60 ml] extra-virgin olive oil

1 Tbsp kosher salt

1½ in [4 cm] thick bone-in rib-eye steak (about 1½ lb [680 g]), at room temperature

1 Tbsp freshly ground black pepper

BEANS

2 Tbsp extra-virgin olive oil

1 tsp minced garlic

8 oz [230 g] canned navy beans, drained

3 fresh rosemary sprigs

1 tsp kosher salt

½ tsp freshly ground black pepper

1. Preheat the oven to 300°F [150°C].

2. To make the steak, heat a medium cast-iron skillet over high heat, add 2 Tbsp of the olive oil, and continue heating until the oil begins to smoke.

3. Sprinkle ½ Tbsp of the salt on one side of the steak and place the steak in the pan salt-side down. Sear for 3 minutes. Lightly oil the top of the steak with the remaining 2 Tbsp of olive oil, sprinkle the remaining ½ Tbsp of salt on top, and then gently flip the steak to sear the other side for 2 minutes. Transfer the hot skillet to the oven and cook for 10 minutes. Check the internal temperature of the steak; it should be about 110°F [43°C] for medium-rare. If you like your steak more cooked, let the internal temperature rise to 120°F to 130°F [50°C to 54°C]. We always err on the side of undercooking because you can always cook it a bit more. The temperature of the steak will continue to rise as it rests.

4. Remove the skillet from the oven, cover with aluminum foil, and let the steak rest for 10 minutes. Top with freshly ground black pepper. Cut the steak against the grain. The meat closest to the bone will be the most flavorful.

5. While the steak rests, prepare the beans. In a medium saucepan over medium-low heat, warm the olive oil. Add the garlic and sauté until fragrant and soft, about 2 minutes. Add the beans, rosemary, salt, and pepper. Stir to coat, cover, and cook for 10 minutes.

6. To serve, divide the beans between two plates. Top with slices of steak and garnish with a sprig of the cooked rosemary. Leftovers can be stored in an airtight container in the refrigerator for 2 to 3 days.

SWEET
SWEET
SWEET
SWEET
SWEET

SWEET
SWEET
SWEET
SWEET
SWEET

Tahitian Vanilla–Cardamom Sheep Milk Ice Cream

While Toluma Farms is famous for its Tomales Farmstead Creamery cheeses, we couldn't in good conscience write this cookbook without sharing another delightful use for the glorious sheep milk we produce on the farm. Sheep milk ice cream is hard to find in the market, but if you're able to get your hands on raw sheep milk, we highly recommend making your own. (Note that raw milk may pose a risk of illness, so please purchase from trusted sources.) We also reduce the sugar content in this ice cream and play up the aromatic flavors of vanilla and cardamom. The result is a very light and creamy dessert that you can customize endlessly throughout the year.

MAKES 3 PINTS [1.4 L]

2 cups [475 ml] heavy cream

1 cup [240 ml] whole sheep milk

½ cup [100 g] superfine sugar

⅛ tsp fine sea salt

6 egg yolks

10 cardamom pods

2 Tahitian vanilla beans

1. Freeze the ice cream maker insert according to the manufacturer's instructions.

2. Crush seven of the cardamom pods to remove the seeds. Grind the seeds in a mortar or clean spice grinder. Set aside. Gently crush the remaining three pods and set aside. Using a paring knife, gently cut the vanilla beans open lengthwise to reveal the seeds. Using the back of the knife, scrape the seeds from the pods. Set aside the seeds and the pods.

3. In a large, heavy-bottom saucepan over medium heat, simmer the cream, sheep milk, sugar, and salt until completely dissolved and steaming but not boiling. Remove from the heat and pour about one-third of the milk mixture into a measuring cup.

4. In a medium bowl, whisk the egg yolks until creamy and smooth. Still whisking, slowly pour in the one-third portion of reserved warm milk mixture from the measuring cup to temper the yolks. Return the pot of milk mixture over low heat and slowly pour in the tempered yolk mixture, stirring constantly for about 5 minutes.

5. Add the ground cardamom, cardamom pods, vanilla seeds, and vanilla pods to the mixture, and continue cooking over low heat, stirring occasionally, until the custard mixture reads 170°F [76°C] on an instant thermometer and is thick enough to coat the back of a wooden spoon.

6. Set a fine-mesh sieve over a large bowl and strain the mixture into the bowl to capture any lumps. Let the mixture cool, then place plastic wrap directly on the surface to prevent a skin from forming. Refrigerate for at least 5 hours and up to 12 hours.

7. Churn the ice cream according to the manufacturer's instructions. It may need to churn longer than traditional cow's milk ice cream. When it's done, the texture should be light, creamy, and fluffy, nearly blooming out of the top of the ice cream maker. The ice cream can be stored in an airtight container in the freezer for up to 1 week.

Mid-August Blackberry Corn Muffins

Wild blackberry season gets into full swing in Marin County starting in early August. By the middle of month, blackberries are dark, sweet, and practically spilling off the bushes on the farm and along the country roads. Every evening walk is an opportunity to harvest. Blackberry jam is my absolute favorite way to enjoy this treat, but blackberry corn muffins are a close second. The corn flour and flaxseed add a mild earthiness and chewy texture that make these perfect any time of day.

MAKES 12 MUFFINS

4 cups [about 480 g] fresh blackberries

⅓ cup [65 g] granulated sugar, plus 2 Tbsp

2 large eggs

¾ cup [180 g] sour cream

½ cup [110 g] unsalted butter, melted

¼ cup [60 ml] canola oil

1 cup [140 g] all-purpose flour

½ cup [70 g] corn flour or finely ground cornmeal

½ cup [45 g] ground flaxseed

2 tsp baking powder

½ tsp kosher salt

1. Preheat the oven to 350°F [180°C] and grease a 12-cup muffin tin with butter or canola oil.

2. In a shallow bowl, sprinkle the blackberries with 2 Tbsp of the sugar, then lightly press the berries with the back of a fork to macerate; set aside.

3. In a small mixing bowl, whisk together the eggs, the remaining ⅓ cup [65 g] sugar, the sour cream, butter, and canola oil; set aside.

4. In a large mixing bowl, whisk together the all-purpose flour, corn flour, flaxseed, baking powder, and salt. Slowly add the wet ingredients to the dry and mix until no lumps remain. Gently fold in half of the berries, being careful not to crush them. Divide the batter evenly among the muffin cups and top each cup with the remaining berries.

5. Bake for 15 to 18 minutes until the edges are golden and a toothpick comes out clean when inserted into the center of a muffin. Let cool in the muffin tin for 10 minutes, then transfer to a wire rack to finish cooling. The muffins can be stored in an airtight container at room temperature for up to 5 days.

Californios Banana Cachapa

This recipe is contributed by chef Emilie Van Dyke and executive chef and owner Val M. Cantu of Californios.

This recipe is a very delicate cachapa, a traditional Venezuelan dish. While not particularly easy to make, it highlights the cheese beautifully. The dough has no gluten, so it has almost no structure, which makes it difficult to cook properly, but it also allows it to be so light and ethereal. The trick is to cook it over a very even heat, allow the bottom layer to become golden brown, and then flip it. We hope this helps in the preparation because we think it is worth it.

MAKES 4 SERVINGS

- 1 egg
- 3 Tbsp whole milk
- 1 tsp agave syrup
- 1 small (10 oz [280 g]) banana
- ½ cup [50 g] corn flour (we like Harina P.A.N.)
- ½ tsp kosher salt
- 1 Tbsp melted cultured butter
- Clarified butter, for greasing the pan
- 6 oz [170 g] fresh goat cheese (like our Liwa)
- Orange blossom honey (we like Marshall's Farm)
- Hoja santa leaves (optional)
- Fleur de sel

1. In a blender, add the egg, milk, agave, banana, corn flour, and salt, taking care to add the dry ingredients last, and blend on medium-high speed until smooth. With the blender running, slowly stream in the melted butter.

2. Heat a 9 in [23 cm] nonstick pan over medium-low heat. Brush the pan with clarified butter and pour ¼ cup [60 ml] of the batter onto the pan. Cook for 1½ to 2 minutes per side, turning the heat to low as needed so that the middle cooks without burning the outsides. Just like with traditional pancakes, look for bubbles to appear on the surface as a good indication that the cachapas are ready to flip.

3. Once flipped, allow the pancake to cook for an additional 30 to 45 seconds. Cachapas, unlike traditional pancakes, are noted for their custard-like texture, so it is important not to overcook them once they're flipped. Remove the cachapa from the heat and allow to cool for a few minutes so that the cheese doesn't melt when they are being filled. Repeat with the remaining batter as the cachapas cool.

4. When cool enough to handle, spread half of each cachapa with the cheese, a drizzle of honey, and hoja santa leaves (if using), and finish with a bit of fleur de sel. Fold so the filling is sandwiched inside the pancake and serve immediately while warm. Much like pancakes, these are best enjoyed as soon as they are prepared.

THE *Grace* OF *Goats*

TERRY GAMBLE BOYER,
SONOMA COUNTY RESIDENT

We have always expected fire. It's a fact of living in California. We have droughts. We have heat. I live in Sonoma. Wine country. Cheese. Every fall, around the time the grapes are harvested, we have fire season.

Our house sits in a field that was previously a cow pasture. We call it Caldera Farms because it mostly grows rock—the hard, handsome basalt spewed millennia ago from the same volcanoes that created the hot springs in Calistoga. If we were to grow vines, we would have to rip the rock from our fields—something we've chosen not to do, opting instead for the moonscape scenery and the plentiful grazing land.

For the last dozen years, we've brought in hundreds of sheep and goats. In 2017, after five years of drought, we had a season of deluges that led to ostentatious grasses, tree canopies, ferns, madrone, manzanita, and, of course, poison oak. We had two goals for the sheep and goats' intensive grazing: First, to improve the health of the soils, and second, to give us a firebreak. At night, a Great Pyrenees named Fluff marked and barked, discouraging coyotes and mountain lions alike. A deranged-looking border collie scooted the herd during the day, ably keeping the sheep contained, less effectively so the goats.

The first kid born in March of 2017 was black and white—I wanted to name it Oreo, but my husband insisted on Hydrox. In August, Hydrox died of pneumonia. We had told ourselves not to get attached. Don't name your meal. Let them do their job. We have a mutually beneficial symbiosis in which our grasses are cut down and the animals are fattened up. But we're not really farmers or ranchers, so in the end, we do get attached.

Fire is conceptual until it happens. We'd taken precautions. The goats and sheep were themselves a precaution. Yes, they were picturesque—but foremost was their utilitarian purpose: the chomping of the grasses in a dry, dry place.

When our neighbors called after midnight on October 7, 2017, I jumped up and got dressed, ran to retrieve my computer, and dragged the pump and hose to the pool so that firefighters might defend the house. From the orange glow in the sky, I knew I wouldn't be sticking around. This wasn't a little grass fire. This was massive.

There is a sensation you get when something unfathomable comes at you. Maybe it's the adrenaline that forces a clarity of purpose (Survive! Survive!); that numbs any fear, horror, or grief that might interfere with the objective to get the hell out. The first thought is, "Is this really happening?" followed by, "It is!"

Some of my decisions were good (the pool, the computer). Some were futile (spray the blue oak outside the bedroom with the garden hose), and some were ridiculous (I put on lipstick before I left the house). Did I mention the sky was orange?

People perished in that year's fires. Billions of dollars went up in smoke. We were the lucky ones. The animals and humans got out. Our house stood. To this day, we still have our land grazed. We climb up trees, eliminate "ladder fuels," and keep a go bag. Some of our non-California friends think we're crazy, but in this time of climate change, we prefer to live in a place we love and invest however we can to protect it, depending in large part on the grace of sheep and goats. It's still hard not to get attached.

Liwa Basque Cheesecake with Honeyed Nectarines and Nasturtiums

We couldn't write a cheese-centric cookbook without featuring a cheesecake! This classic Basque-style cheesecake with a caramelized toasted top is forgiving enough to be a slam dunk even for cooks who are intimidated by water baths. While traditional cheesecakes have crumbly crusts and bake for a long time, the Basque-style cheesecake is completely crustless and bakes at a very high temperature for a shorter period of time, leaving a dark and toasty scented top. When you peel away the parchment after baking, you're left with a gorgeous ripple pattern and a creamy and light interior. We like to feature the Liwa in this cake and serve it in the summer with stone fruit.

MAKES 6 TO 8 SERVINGS

10½ oz [300 g] fresh goat cheese (like our Liwa), at room temperature

1¼ cups [300 g] Jersey milk cream cheese, at room temperature

1 cup [200 g] granulated sugar

4 large eggs, at room temperature

1¼ cups [300 ml] heavy cream

1 tsp vanilla extract

5 Tbsp [40 g] all-purpose flour

Pinch of kosher salt

2 ripe nectarines, sliced into thin wedges

3 Tbsp wildflower honey

Nasturtiums, for garnish

1. Preheat the oven to 400°F [200°C]. Lightly grease the inside of a 9 in [23 cm] springform pan. Crinkle up 2 rectangles of parchment paper in your hands to make them more flexible, then place them in the pan so they form a cross and hang over the top of the pan by a few inches.

2. In a large mixing bowl or in a stand mixer fitted with a paddle attachment on medium-high speed, beat the fresh goat cheese and cream cheese together until light and fluffy. Add the sugar slowly and beat on medium speed to incorporate. Beat in the eggs at medium speed one at a time, only adding the next egg after the previous one is fully incorporated. Add the heavy cream slowly, followed by the vanilla, and beat on low speed just to combine.

3. Sift in the flour and salt and gently fold it into the cheese mixture. Pour the mixture into the prepared pan and transfer to the center rack of the oven. Bake for 50 min to 1 hour until the top appears dark and the cheesecake has puffed above the rim. Let cool at room temperature for 1 hour, then transfer to the refrigerator for 1 hour.

4. While the cake chills, place the nectarines in a saucepan over low heat with the honey. Stir until the nectarines are warm and the honey has melted and coated the fruit. Remove from the heat.

5. After 1 hour in the refrigerator, the cake will have sunk into the pan. Carefully remove the cake from the springform pan and peel the paper away. Top with the nectarines and garnish with nasturtiums.

6. To serve, slice into wedges. The cheesecake can be stored in an airtight container in the refrigerator for up to 3 days, but we recommend eating this dish the same day it's made.

Rhubarb Crumble Cake

This rhubarb cake is my birthday tradition. In the late spring and early days of summer, it's time to harvest the massive leaves and stalks and look for new uses for rhubarb. Rhubarb has a short season, but we often stock up and freeze it for use all summer long. We love this cake because there is something for everyone: a gleaming pink rhubarb top, a traditional and delicate vanilla sponge, and a crumble on the bottom for lovely texture and sweetness. This is technically an upside-down cake, where the rhubarb is cooked at the bottom of the cake, the crumble is piled on top, and then the whole thing is inverted on a cake stand once it's taken out of the oven. We like to serve each slice on its side so that every texture and layer is presented.

MAKES 8 SERVINGS

CRUMB TOPPING

1 cup [140 g] all-purpose flour

½ cup [100 g] granulated sugar

½ cup [110 g] unsalted butter, melted

¼ tsp kosher salt

CAKE

1 lb [455 g] rhubarb, trimmed and cut into ½ in [13 mm] pieces

1¾ cups [350 g] granulated sugar

¾ cup [165 g] unsalted butter, at room temperature, plus more for buttering the pan

1½ cups [210 g] all-purpose flour

1½ tsp baking powder

½ tsp kosher salt

½ tsp finely grated lemon zest, plus more for garnish

2 large eggs

1 cup [240 g] sour cream or whole milk yogurt

1. Preheat the oven to 350°F [180°C].

2. To make the crumb topping, in a medium bowl, combine the flour, sugar, butter, and salt until the mixture forms the texture of wet sand.

3. To make the cake, in a small saucepan, add the rhubarb, ¾ cup [150 g] of the sugar, and a splash of water. Place over medium heat and stir until the sugar dissolves and the rhubarb begins to soften and release its juices, about 8 to 10 minutes. Remove from the heat to cool.

4. Butter a 9 in [23 cm] round cake pan (2 in [5 cm] deep) and line the bottom with parchment paper. Dice ¼ cup [55 g] of the butter into small cubes and sprinkle it around the bottom of the pan. Pour the rhubarb mixture evenly over the butter; set aside.

5. In a large bowl, whisk together the flour, baking powder, and salt; set aside.

6. In a stand mixer or using a hand mixer, beat the remaining ½ cup [110 g] of butter and 1 cup [200 g] of sugar on medium speed until pale and fluffy. Add the lemon zest, followed by the eggs, 1 at a time, beating until incorporated and scraping down the sides of the bowl. With the mixer running on low speed, slowly add the flour mixture in 4 additions, alternating with the sour cream, until the batter is completely smooth and no lumps or spots of flour remain. Spread the batter evenly over the rhubarb. Add the crumble topping evenly over the batter.

7. Bake until a toothpick inserted into the center comes out clean and the top springs back when touched, about 1 hour. Let cool for 10 to 20 minutes before carefully inverting the cake onto a stand and removing the cake pan. To serve, finely grate lemon zest over the top and cut into slices. The cake can be stored in an airtight container in the refrigerator for up to 5 days, but we recommend eating this dish hot out of the oven.

AUTUMN

Friendsgiving: Our Community of Women-Run Farms

According to Oxfam's website, "about 80 percent of the world's food is produced by small-scale farming and women make up on average 43 percent of this agricultural labor in developing countries. They are the majority in some countries. In South Asia, more than two thirds of employed women work in agriculture. In eastern Africa, over half of farmers are women." According to the census of agriculture, California women-run farming operations represent 37 percent of all the producers in the state. These numbers are expected to continue growing since women are outpacing men in 4-H programs and agricultural education.

In the early 1990s (many years before we had a farm), I was fortunate to visit women-owned and -operated micro farms in Vietnam and Cuba. It was striking to see the good these women cultivated in their communities. Women farmers have continued to inspire me over the decades. They are not just growing food for their family; their passion is to uplift entire communities.

My friend Adriana Silva is also a first-generation farmer; she and her husband own and operate Tomatero Farms in Watsonville, California. Standing in the middle of her farm with the Pacific coast in the background, she told me, "People should be paid well for feeding people. Employees should have healthcare, and organic food should be accessible to all populations." Farmworkers historically—and too often still today—are paid low wages, live in unsafe housing conditions, and work long, arduous hours. Farmworkers who are undocumented usually don't have a voice or anyone in their corner. Low wages, lack of access to healthy food, and the looming threat of exploitation and deportation all make farm work extraordinarily difficult for an estimated 75 percent of the country's farmworkers. Adriana seeks to remedy this at her Watsonville farm, saying, "We treat our workers well. You can tell when people want to work for you—it's because they feel like a part of the farm."

For many years, we'd see Moira Kuhn of Marin Roots Farms working on her farm and selling her organic produce at the farmers' market with her baby on her back and her toddler at her side. She really was juggling it all! Moira tells me what motivates her as a farmer: "What it boils down to is, what do we want to feed our family? Our customers, home shoppers and restaurants, are all putting trust in us to grow items to feed their families that are the same as we'd feed our family."

Priscilla Lucero and her husband, Curtis, own and run a 20-acre ranch in Galt, in the Sacramento Valley, where they are creating a sustainable, biodiverse farm that encompasses the entire ecosystem. To maintain soil fertility, they rotate crops, plant cover crops, and apply compost. At the farm dinners they host, guests can learn about where the food they're enjoying comes from. Priscilla reflects, "Farming is empowering because not only are you feeding yourself but you're feeding and educating others as well. We never would have dreamed that we'd be able to have this type of life. Even if you just start out with one acre and grow from that, that's what our goal is, to keep the next generation looking forward."

I have had the pleasure of sitting on the Farmers' Market Board of Directors (Agricultural Institute of Marin) with Moira, Priscilla, Adriana, and many more women farmers, and I am in constant awe that they not only get up at the crack of dawn and work long, physically demanding hours, but still find time to be actively engaged in their communities. When school starts back up each fall, all of these women begin hosting school groups out to their farms to demonstrate firsthand the importance of regenerative farming, to enable kids to taste the deliciousness of a strawberry picked from the field, and to show them that absolutely anyone can become a farmer.

TO START
TO START
TO START
TO START
TO START

TO START
TO START
TO START
TO START
TO START

Warm Olives with Atika, Herbs, and Lemon Zest

When the days get shorter and the evenings are cool, the long dinners al fresco come to an end. However, I still love to gather outside with some candles, cozy throw blankets, and a spread of finger foods everyone can help themselves to. Warm roasted olives are a fantastic starter, and when they're paired with fresh bread, good cheese, other nibbles, and plenty of good wine, you may end up wanting to linger long enough to watch the stars come out.

MAKES 4 SERVINGS

2 cups [380 g] mixed black and green olives, pitted

4 to 6 garlic cloves, skins left on

2 Tbsp extra-virgin olive oil

2 Tbsp finely chopped fresh thyme

1 Tbsp finely chopped fresh rosemary

¼ tsp crushed fennel seeds

¼ tsp crushed coriander seeds

½ tsp freshly ground black pepper

1 tsp finely grated lemon zest

2 oz [60 g] Parmesan-style cheese (like our Atika), shaved

1. Preheat the oven to 450°F [230°C].

2. In a medium bowl, toss together the olives, garlic, olive oil, thyme, rosemary, fennel, and coriander to coat, then spread on a small baking sheet or in a cast-iron skillet. Roast for 15 to 20 minutes.

3. To serve, transfer the warmed olives to a serving bowl, toss with the pepper and lemon zest, then top with cheese shavings. The olives can be stored in an airtight container in the refrigerator for up to 1 week.

Goat Cheese Tatin

This recipe is contributed by Roland Passot, co-owner and chef at Left Bank Brasserie.

Award-winning French chef and restaurateur Roland Passot has been a staple in the Bay Area since the late 1980s. Since running La Folie in San Francisco and Left Bank Brasserie in Larkspur and Menlo Park, Roland has made a tradition of visiting the Marin Farmers' Markets every Thursday and Sunday to buy Tomales Farmstead Creamery goat cheese and has formed a close friendship with Tamara and David. In Roland's own words:

"Who would have guessed that this man at the farm stand selling goat cheese was a cancer surgeon! David was there every Sunday, and we talked about cancer research. I have a lot of admiration for the guy—between his research, working on the farm and milking the goats, making his bagels—I don't know how he does it. He used to deliver goat cheese to my restaurant in his scrubs."

Jessica was fortunate to be invited by Roland to his home to recreate his signature Liwa dish at La Folie, the Goat Cheese Tatin. This is a layered vegetarian entree with a five-star presentation. Though each layer is simple to make, it's best to prepare the elements for each layer in advance and give the tart time to rest in the refrigerator. The tatin is assembled in a ring form, seared, and lightly baked just before serving. Don't be intimidated by the number of steps; the result is worth it.

MAKES 6 SERVINGS

CHEESE MIXTURE

1 egg white

1 bunch chives, chopped

¼ cup [10 g] chopped fresh thyme

8 oz [230 g] fresh goat cheese (like our Liwa) or fresh chèvre, at room temperature

Kosher salt

Freshly ground black pepper

2 cups [120 g] panko, blended to a fine texture

ARTICHOKES

6 medium artichokes or one 8 oz [230 g] can artichoke bottoms, diced

2 Tbsp extra-virgin olive oil

3 oz [85 g] sun-dried tomatoes, chopped

1 Tbsp minced garlic

1 Tbsp chopped fresh thyme leaves

½ tsp kosher salt

¼ tsp freshly ground black pepper

continued

1. To make the cheese mixture, in a medium bowl, whisk the egg white to a milky consistency, then add the chives and thyme and whisk to combine. Fold in the goat cheese and season with salt and pepper. Add the panko and mix until it reaches a dough-like consistency. Season again. Set aside.

2. To prepare the artichokes, if you're using fresh artichokes, take the bottom leaves off, then remove the stem and fibers. Remove all the outer green leaves from the artichoke with a paring knife and shave down the base until you have a clean artichoke bottom. Dice the artichoke bottoms.

3. In a medium skillet over medium-high heat, warm the olive oil, then add the artichokes and sauté for 5 to 8 minutes. Add the sun-dried tomatoes, garlic, thyme, salt, and pepper and sauté until caramelized, about 8 to 10 minutes. Set aside.

continued

4. To make the tomato confit, preheat the oven to 350°F [180°C].

5. Place all of the halved tomatoes skin-side up in a baking dish and add the olive oil, thyme sprigs, garlic, and peppercorns. Bake until the skin is peeling away from the tomatoes, about 20 minutes. Set aside to cool and lower the oven temperature to 200°F [95°C]. Set 4 of the garlic cloves aside to be used in the sauce vierge.

6. Once the tomatoes have cooled, gently remove the skins, placing the skins on a wire rack set over a baking sheet. Bake the skins for 30 minutes to dehydrate and crisp.

7. To make the eggplant mixture, in a medium skillet over medium-high heat, warm the olive oil. Add the eggplant and 2 Tbsp of the minced garlic and sauté for 5 to 8 minutes. Remove from the heat and add the parsley. Season with the salt and pepper and add the lemon juice and sherry vinegar. Set aside.

8. To make the mushroom layer, preheat the oven to 375°F [190°C]. Place the mushrooms in a baking dish with the tops facing up and drizzle with the olive oil. Add the garlic, thyme, rosemary, salt, and pepper. Bake for 10 to 12 minutes until the mushrooms are tender. Set aside.

continued

TOMATO CONFIT

4 Roma tomatoes, halved, cored, and seed pulp reserved

1¼ cups [300 ml] extra-virgin olive oil

10 fresh thyme sprigs

6 to 8 whole garlic cloves, skins left on

1 tsp whole black peppercorns

EGGPLANT MIXTURE

3 Tbsp extra-virgin olive oil

4 Japanese eggplants, finely diced

5 garlic cloves plus 3 Tbsp finely minced garlic

1 cup [40 g] chopped fresh parsley

½ tsp kosher salt

¼ tsp freshly ground black pepper

Juice of 1 lemon

1 tsp sherry vinegar

MUSHROOM LAYER

6 portobello mushrooms, stems and gills removed

3 Tbsp extra-virgin olive oil

2 whole garlic cloves

2 fresh thyme sprigs

Leaves of one 5 in [12 cm] rosemary sprig, roughly chopped

1 tsp kosher salt

½ tsp freshly ground black pepper

continued

9. To make the sauce vierge, remove the skins from the garlic cloves that were used in the tomato confit. In a blender, add the tomatoes, garlic, basil, olive oil, vinegar, salt, and pepper. Blend well to a smooth consistency. Set a fine-mesh sieve over a bowl and strain the mixture to remove the skin and seeds. Set aside.

10. To prepare the zucchini, in a large sauté pan over medium-high heat, warm the olive oil. Sauté the zucchini until slightly translucent and warmed through but not brown, about 2 to 3 minutes. Transfer the zucchini to a paper towel–lined plate to drain.

11. To assemble the tatins, coat the inside of six 4 in [10 cm] ring molds with olive oil. Roll approximately 2½ oz [70 g] of the cheese mixture into a ball, then place it inside a ring. Flatten the cheese until it sits flush against all sides of the ring. Add 2 Tbsp of the eggplant mixture and press it gently into the cheese. Add 1 Tbsp of the tomato confit on top and gently press. Add 2 Tbsp of the artichoke mixture. Lastly, add a portobello mushroom on top and press gently. Repeat with the remaining filling to assemble the remaining tatins. Let rest for at least 3 hours in the refrigerator.

12. Preheat the oven to 375°F [190°C].

13. Remove the tatins from the refrigerator and gently press the cheese side of each tatin into the dish of reserved panko until it is fully coated.

14. To finish, in a 10 in [25 cm] cast-iron skillet or nonstick pan over medium-high heat, warm the olive oil until it shimmers. Place one of the tatins in the pan (bread crumb–side down). Let it color for a few minutes until golden brown and a slightly dark crust forms. It should look like the top of a crème brûlée.

15. Transfer the tatin to a baking sheet, portobello-side down. Repeat for the remaining tatins. Bake for 15 minutes.

16. To serve, divide the zucchini and sauce vierge among 6 serving plates. Slide a paring knife along the inner edges of each ring mold to loosen the tatin and unmold it very delicately. Transfer the tatin onto the zucchini and sauce vierge. Garnish each with basil, dried tomato skin, and kalamata olives and serve. If you would like to prepare the tatins ahead, they can be assembled and refrigerated in their ring molds for up to 4 hours before frying and serving. Once plated, they should be enjoyed immediately.

SAUCE VIERGE

4 Roma tomatoes, cored and coarsely chopped

6 to 10 fresh basil leaves

1½ cups [360 ml] extra-virgin olive oil

⅓ cup [80 ml] sherry vinegar

1 tsp kosher salt

½ tsp freshly ground black pepper

ZUCCHINI MIXTURE

2 Tbsp extra-virgin olive oil

3 zucchini, sliced into ⅛ in [4 mm] rounds

TO FINISH

2 Tbsp extra-virgin olive oil

½ cup [30 g] panko, blended to a fine texture

6 fresh basil leaves

12 to 18 Kalamata olives, pitted and halved

FARMERS *Are* CONSERVATIONISTS

LILY VERDONE, EXECUTIVE DIRECTOR, MARIN AGRICULTURAL LAND TRUST (MALT)

Farmers and ranchers are vital to our way of life. They grow our food and fiber, are the backbone of our local economies, and are among our greatest conservation allies. Supporting farmers and ranchers with collaborative, science-based, regenerative approaches ensures that we can feed a growing population while maintaining clean and abundant water supplies, healthy lands, and a stable climate.

The Marin Agricultural Land Trust—the first agricultural land trust in the country—started in 1980 knowing that to ensure a resilient local food shed, we need to also preserve the land where food is grown. At our heart, we are a community-based organization—growing our impact, sharing stories, and supporting our local agricultural economy.

Our mission is still strong today because food brings people together, and those who are good at getting people together help build a shared future that is better for all. When we invest in people, place, and economy, it is a win-win.

Roasted Harissa Root Vegetables with Koto'la and Toasted Pumpkin Seeds

This recipe is dedicated to Tina Trevino.

We couldn't imagine a Toluma Farms cookbook without a recipe dedicated to Tina Trevino, the farm's herd manager from 2019 to 2024. She showed me how to look after the goats and sheep, work in the barn, and build caring connections with each animal in the herd. I am endlessly inspired by her.

The first dish I ever made for Tina was a cassoulet, and her immediate feedback was, "I love the giant carrots, Jessica! More big vegetables, please." The very next day, I developed this recipe for her—an entire dish dedicated to big chunks of roasted vegetables with layers of flavors and textures that turn something healthy into an incredibly addictive treat.

MAKES 6 SERVINGS

¼ cup [10 g] roughly chopped fresh thyme leaves

3 Tbsp extra-virgin olive oil

1 Tbsp harissa

1 Tbsp kosher salt

1 lb [455 g] radishes, halved

1 lb [455 g] carrots, peeled and cut into 2 to 3 in [5 to 7.5 cm] diagonal chunks

8 oz [230 g] parsnips, peeled and cut into 2 to 3 in [5 to 7.5 cm] diagonal chunks

2 fennel bulbs, sliced

6 whole garlic cloves, skins left on

¼ cup [35 g] pumpkin seeds

½ cup [120 g] crumbled feta-style cheese (like our Koto'la)

¼ cup [10 g] roughly chopped fresh parsley

Freshly ground black pepper

1. Preheat the oven to 425°F [220°C] and place a large, unlined baking sheet in the oven to preheat.

2. In a large bowl, mix together the thyme, olive oil, harissa, and salt. Add the radishes, carrots, parsnips, fennel, and garlic and toss to coat.

3. Carefully remove the hot baking sheet from the oven and evenly spread the vegetables on the sheet. Avoid crowding and consider using two baking sheets if needed. Roast for 30 to 45 minutes, checking halfway through to toss the vegetables and turn the pan for even roasting.

4. Meanwhile, in a small, dry skillet over medium-low heat, toast the pumpkin seeds for about 5 minutes until lightly browned and fragrant.

5. Once the vegetables are out of the oven and the garlic is cool enough to handle, carefully remove the garlic from the skins by squeezing them out. To serve, toss everything together with the garlic and transfer the vegetables to a serving platter. Top with the pumpkin seeds, crumbled cheese, parsley, and pepper and serve. Leftovers can be stored in an airtight container in the refrigerator for up to 3 days.

Goat Cheese Croque Monsieur Canapé

Whenever we order a croque monsieur sandwich, it's so rich that we can enjoy only the first few bites before setting it aside for later. This led us to the perfect canapé to serve with cocktails: a goat cheese croque monsieur! One bite-size square of rich brioche with ham and Gruyère to balance a sharp drink or glass of Champagne—the perfect pairing for your next party.

MAKES 8 SERVINGS

2 Tbsp Dijon mustard

3 Tbsp fresh goat cheese (like our Liwa)

8 brioche slices, about ½ in [13 mm] thick

8 oz [230 g] thinly sliced cured ham

8 oz [230 g] thinly sliced aged Gruyère cheese

¼ cup [55 g] unsalted butter, at room temperature

1. Preheat the oven to 400°F [200°C] and line a baking sheet with foil or parchment paper.

2. In a small bowl, mix the mustard and fresh goat cheese together to form a pale yellow spread. Spread a very thin layer of the mixture across each slice of brioche.

3. On half of the bread slices, layer two thin slices of ham topped with two thin slices of Gruyère. Top with the remaining bread slices, then butter the outside of the sandwiches. Place the four sandwiches on the baking sheet and bake for 10 minutes until the bread is toasted and the cheese is melted.

4. To serve, transfer the sandwiches to a cutting board. Trim the crusts off each sandwich and cut into 1 in [2.5 cm] squares. Spear each square with a toothpick and transfer to a platter. Leftovers can be stored in an airtight container in the refrigerator for up to 1 day.

Toasted Liwa Muffin with Dill and Honey

This is the dish that converted Jessica from a dill hater to a dill lover. There is a magical alchemy between the honey and the Liwa that softens the pungent nose on the dill and makes you want to top each slice with more. She first tasted a dish similar to this as an appetizer at a wine bar in Martha's Vineyard and immediately tried to recreate it upon returning home. We recommend enjoying this with a light salad for lunch or slicing it into small wedges to serve as a starter.

MAKES 2 SERVINGS

- 1 oz fresh goat cheese (like our Liwa) or fresh chèvre
- ½ tsp fine salt
- 3 Tbsp roughly chopped fresh dill leaves
- 2 Tbsp wildflower honey, warmed until runny
- 2 English muffins, halved and toasted
- ¼ tsp freshly ground black pepper

1. In a small bowl, combine the cheese, salt, 1 Tbsp of the dill, and 1 Tbsp of the honey. Spread evenly across the four slices of toasted English muffin. Top each slice with a sprinkle of the remaining dill, a light drizzle of the remaining honey, and the pepper. Serve immediately.

Potato Leek Soup with Smoked Pepper and Liwa Crema

In my family, it was tradition for my noni (grandmother) to serve potato leek soup for lunch on Thanksgiving. It's rich, creamy, and satisfying but leaves you well positioned to take on a large meal later in the day. It was the only time of year I had it, and to this day, it marks the start of the holiday season for me. I love serving it garnished with creamy goat cheese, spicy black pepper, and toasted pine nuts or pumpkin seeds.

MAKES 6 TO 8 SERVINGS

3 Tbsp unsalted butter

1 Tbsp extra-virgin olive oil

2 leeks, white and light green parts, finely chopped

1 tsp kosher salt

6 large new potatoes or creamer potatoes, diced

2 bay leaves

4 fresh thyme sprigs

2 fresh rosemary sprigs

6 cups [1.4 L] chicken stock, preferably homemade

6 oz [170 g] fresh goat cheese (like our Liwa), plus more for garnish

½ cup [120 ml] heavy cream

1 Tbsp black peppercorns

1. In a large soup pot over medium-low heat, melt the butter until the foaming has subsided and it begins to brown and is fragrant, about 2 to 3 minutes. Add the olive oil to slow the browning, followed by the leeks and the salt. Stir to coat the leeks and sauté until softened, about 5 minutes. Add the potatoes and stir.

2. Create a bundle with the bay leaves, thyme, and rosemary and tie it together with cotton twine to form a bouquet garni. Add it to the pot and cover to let the potatoes soften, about 5 minutes. Add the chicken stock and turn the heat up to medium. Cover and simmer for 15 to 20 minutes until the potatoes are soft and can be easily pierced with a knife.

3. In a medium bowl, stir the cheese into the heavy cream, then add the mixture to the soup. Remove from the heat and, using an immersion blender, blend the soup until silky and smooth.

4. In a small, dry skillet over medium heat, toast the peppercorns for 3 to 5 minutes. Transfer to a rimmed wooden cutting board and carefully crush using the bottom of the skillet.

5. To serve, portion the soup into shallow bowls and top with freshly crushed peppercorns and a dollop of cheese. The soup can be stored in an airtight container in the refrigerator for up to 5 days.

No, Goats Don't Eat TIN CANS

TINA TREVINO, TOLUMA FARMS HERD MANAGER, 2020–2024

One of the biggest misconceptions about goats is that they will eat anything. After working with goats for more than five years, I can confidently say this is not the case. They actually can be exceptionally picky eaters.

A few of our goats in particular—Peggy, Oreo, and Krakatoa—will methodically sort through their grain to eat all the corn bits and leave the pellets behind. I can't help but be reminded of children who want to eat only the marshmallows out of a bowl of Lucky Charms.

Based on my experience, I believe the misconception comes from the fact that goats (also like small children) want to put everything in their mouths. Goats are incredibly curious creatures who explore the world around them primarily through their taste buds. This tendency can be viewed as either cute (watching goat kids nibble on their mama's goatee) or a working hazard (losing the end of my braid to a curious nibbler), but its primary function is evolutionary.

Goats are natural browsers, meaning they like to eat plant matter such as bushes, shrubs, vines, and bark, not just grass. With their super-sensitive taste buds, they can usually detect what is safe for them to consume and what might be dangerous or toxic. By tasting plants first, they can avoid a potentially deadly meal. Baby goats learn what's safe to eat by browsing and grazing with their moms and the rest of the herd. They watch the adults, taste what they taste, and eat what they eat, thus acquiring what I can only assume is a very sophisticated goat palate.

I love watching each goat grow into its own unique personality as they explore the world around them. The best lesson they have taught me is to always lead with curiosity.

MAINS
MAINS
MAINS
MAINS
MAINS

MAINS
MAINS
MAINS
MAINS
MAINS

Maple-Crusted Butternut Squash with Liwa

The rich sweetness and smokiness of this roasted butternut squash ensures its place on the table as a main course. While the squash is naturally earthy and sugary, it's the maple drizzle that really sets off the smoky and salty hit from the bacon. Don't fear: The mild and cool Liwa mellows the whole platter once it's ready to serve, and everyone will be going in for seconds.

MAKES 4 TO 6 SERVINGS

1 large butternut squash, peeled, seeded, and cut into ½ in [13 mm] cubes

½ yellow onion, finely chopped

¼ cup extra-virgin olive oil

4 to 5 garlic cloves, minced (reserve 1 Tbsp for drizzle)

4 Tbsp chopped fresh thyme leaves, plus more for garnish (optional)

1 Tbsp kosher salt

2 tsp freshly ground black pepper

4 to 5 bacon strips, chopped into 1 inch [2.5 cm] pieces

¼ cup [60 ml] maple syrup

½ cup [70 g] pumpkin seeds

8 oz [230 g] fresh goat cheese (like our Liwa)

1. Preheat the oven to 425°F (220°C). Line a large baking sheet with parchment paper.
2. Place the butternut squash in a large bowl. Add the onion along with the olive oil, garlic, 3 Tbsp of thyme, salt, and pepper. Mix gently to combine.
3. Spread the squash mixture onto the sheet. Sprinkle the bacon over the top of the mixture and bake for 20 minutes.
4. While the squash bakes, in a small bowl, mix together the maple syrup, pumpkin seeds, and the reserved 1 Tbsp of garlic and 1 Tbsp of thyme.
5. Drizzle the maple mixture over the top of the squash and bake for another 20 minutes.
6. To serve, arrange the squash on a large serving platter and immediately spoon the cheese onto the squash. Garnish with more thyme, if desired. This dish can be stored in an airtight container in the refrigerator for up to 3 days.

Smoky Baked Eggs with Chickpeas and Goat Cheese

This is the perfect thing to make when you want to have breakfast for dinner. The smoked paprika dresses up the eggs and pairs so nicely with a dark red wine. The chickpeas give this dish incredibly satisfying savoriness and texture. The fresh herbal hit of parsley and salty bite of Koto'la at the end round everything out. This is a great dish for date night—or any night.

MAKES 4 SERVINGS

2 Tbsp extra-virgin olive oil

2 Tbsp salted butter

4 garlic cloves, sliced

2 tsp smoked paprika

2 cups [480 ml] canned crushed tomatoes

One 15 oz [425 g] can unsalted chickpeas, rinsed and drained

½ tsp kosher salt

1 Tbsp chopped fresh thyme

6 large eggs

2 oz [55 g] feta-style cheese (like our Koto'la)

½ tsp freshly ground black pepper

3 Tbsp roughly chopped fresh parsley

Sourdough bread, for serving

1. In a 12 in [30.5 cm] skillet over medium-low heat, warm the olive oil and butter. Add the garlic and smoked paprika and cook, stirring, until the garlic is fragrant and beginning to brown, about 2 minutes. Increase the heat to medium and add the tomatoes, chickpeas, salt, and thyme and stir. Adjust the heat to maintain a simmer.

2. Crack an egg into a small bowl, taking care not to break the yolk. Make a well in the sauce using the back of a ladle or a glass, roughly large enough to hold the egg, and slide the egg into the well so that the yolk and most of the white is contained (some white may spread out). Repeat with the remaining eggs, evenly spacing them around the skillet.

3. Sprinkle the cheese evenly over the pan and top with the pepper. Cover the pan to allow the steam to cook the eggs until they reach the desired doneness, 6 to 8 minutes for medium-set where the yolks will be runny and the whites will be glazed over and firm. Remove from the heat.

4. To serve, sprinkle with the parsley and serve with sourdough bread. We recommend eating this dish immediately, but the sauce and chickpeas can be stored in the refrigerator in an airtight container for up to 3 days. Reheat and continue with cooking the eggs.

Ross's Niçoise Salad

This is my father-in-law's favorite salad and might soon be yours as well. If you're able to find sushi-grade ahi tuna, I highly recommend using it here and searing it just enough to form a salty crust while leaving the center a gleaming fuchsia pink. This is an extremely Californian twist on a French classic.

MAKES 4 SERVINGS

DRESSING

3 Tbsp extra-virgin olive oil

1 Tbsp champagne vinegar

1 tsp Dijon mustard

1 tsp minced shallots

SALAD

1 lb [455 g] baby potatoes, sliced into ½ in [13 mm] chunks

3 Tbsp vegetable oil

½ tsp kosher salt, plus more for seasoning

4 large eggs

1 lb [455 g] fresh green beans, ends trimmed

2 or 3 heads crisp Little Gem lettuce, leaves separated from core

1 pint [300 g] cherry or grape tomatoes

½ cup [85 g] Castelvetrano olives

½ tsp roughly chopped fresh dill

2 oz [55 g] fresh basil, julienned

8 oz [230 g] fresh ahi tuna

¼ tsp freshly ground black pepper

1. Preheat the oven to 400°F [200°C] and line a baking sheet with parchment paper.

2. To make the dressing, in a large salad bowl, whisk together the olive oil, vinegar, mustard, and shallots. Set aside.

3. To make the salad, spread the potatoes evenly on the prepared baking sheet. Toss with 1 Tbsp of the vegetable oil and the salt. Bake for 20 to 25 minutes until crisp and browned. Set aside to cool.

4. Bring a large pot of water to a boil and fill a bowl with ice water. Boil the eggs for 6 minutes, then transfer to the bowl to cool. Add the green beans to the pot of water and return to a boil. Boil for 4 minutes, then drain and rinse under cold water.

5. Add the green beans, lettuce, tomatoes, olives, dill, and basil to the salad bowl with the dressing and refrigerate while you sear the tuna.

6. In a 10 in [25 cm] cast-iron pan over high heat, warm the remaining 2 Tbsp of the vegetable oil until very hot and shimmering. Season each side of the tuna with salt and the pepper and place into the hot oil. Sear for about 2 minutes on each side, then transfer to a cutting board and slice into ½ in [13 mm] slices. The fish will be rare in the center.

7. To serve, toss the salad mixture with the dressing and plate in four equal portions. Top each portion with a few slices of ahi. Peel the eggs under cool running water and slice each egg in half lengthwise, placing one egg on each salad. Serve immediately. We recommend eating this dish the same day it's made.

Harvest Lentil Soup with Fried Liwa and Butter

Lentil soup takes many forms across cultures, from Italy to India, but it is always deeply nourishing, satisfying, and somehow feels rich and healthy at the same time. In this recipe, we use green lentils and fresh baby spinach and always finish it with a couple tablespoons of butter. You can top each bowl with a small disk of lightly breaded and fried goat cheese, which will melt beautifully into the hot soup. We suggest serving in large, shallow bowls with a side of warm country bread.

MAKES 4 SERVINGS

LENTIL SOUP

2 Tbsp extra-virgin olive oil, plus more for finishing

1 yellow onion, diced

4 to 6 garlic cloves, minced

2 tsp kosher salt

1 tsp freshly ground black pepper

2 large carrots, diced

2 to 3 celery stalks, diced

2½ cups [500 g] green lentils, rinsed

7 cups [1.7 L] chicken stock

1 Tbsp finely chopped fresh rosemary

4 to 6 fresh thyme sprigs

4 cups [80 g] fresh baby spinach leaves

2 Tbsp salted butter

½ cup [20 g] roughly chopped fresh parsley

FRIED GOAT CHEESE

2 Tbsp extra-virgin olive oil

¼ cup [35 g] finely ground panko

4 oz [115 g] fresh goat cheese (like our Liwa) or fresh chèvre

1. To make the lentil soup, in a large stockpot over medium-high heat, warm the olive oil. Add the onion, garlic, salt, and pepper, stirring occasionally until slightly translucent and fragrant but not brown, about 4 minutes. Add the carrots and celery and stir to coat in the oil. Cover and cook for approximately 5 minutes until the vegetables are fragrant and slightly tender.

2. Stir in the lentils, stock, rosemary, and thyme and lower the heat to medium-low so the soup comes to a very gentle simmer. Cover and simmer for about 1 hour, stirring occasionally. Add the spinach, stir, and cover for another 15 minutes while you make the fried goat cheese.

3. To make the fried goat cheese, in a medium pan over medium-high heat, warm the olive oil. Add the panko to a shallow dish. Divide the goat cheese into 1 oz [30 g] portions. Flatten each portion between your fingers into small disks about ¼ in [6 mm] thick. Lightly dip each disk into the panko, coating all sides, and fry in the hot oil for about 2 minutes per side. Lower the heat if the oil begins to smoke. When done, the disks should resemble the color of crème brûlée—lightly golden with brown, crusted edges.

4. When the soup is done, remove from the heat. Remove and discard the thyme sprigs and stir in the butter until melted. To serve, portion the soup equally into four bowls, then top with the parsley, fried goat cheese, and a drizzle of olive oil to serve. The soup can be stored in an airtight container in the refrigerator for up to 3 days. The goat cheese should be freshly fried before serving.

SWEET
SWEET
SWEET
SWEET
SWEET

SWEET
SWEET
SWEET
SWEET
SWEET

Gravenstein Apple and Marmalade Grilled Cheese

Each year, the farm and creamery participate in the Gravenstein Apple Fair in our neighboring town of Sebastopol. These incredible apples are a slow food heritage fruit that originated in the seventeenth century. The farm planted a small orchard of Gravenstein apples and other heritage apple trees, which we use for our autumn cheese boards, cider, and delicious meals such as this.

We love whipping up this sandwich as a quick afternoon snack for our farm staff. It's fast, it fills them up, and it wakes up the senses. The bitter bite of the marmalade is mellowed by the smooth saltiness of the cheese and crisp sweetness of the apples. Make sure to use crusty sourdough bread and cook this grilled cheese sandwich low and slow in a cast-iron skillet for the perfect melty texture.

MAKES 1 SANDWICH

2 Tbsp salted butter

2 Tbsp orange marmalade

2 thick slices sourdough bread

2 to 3 oz [55 to 85 g] triple cream cheese (like our Teleeka)

1 small Gravenstein or other tangy-sweet, fragrant apple, cut into thin slices

1. In a cast-iron skillet over medium heat, lightly brown the butter.

2. Meanwhile, spread a thin layer of marmalade on one side of each slice of bread. Add one slice to the pan, marmalade-side up. When it begins to crisp, layer on the cheese and apple slices. Top with the remaining slice of bread, marmalade-side down, and gently flip. Lower the heat to low and cook until the cheese is softened, about 3 minutes.

3. Transfer to a cutting board, slice diagonally, and serve immediately. We recommend eating this dish the same day it's made.

Bay Laurel and Fig Clafoutis

While bay leaves are a staple in savory cooking such as stocks, stews, and braises, we love to incorporate the herbal hit of bay laurel in sweet dishes too. It's abundant in Northern California year-round and can be used fresh or dried by steeping in your cooking liquid of choice. A clafoutis is a light dessert that has milk and egg batter puffed in the oven around a generous amount of fruit. In this case, the bay leaves are steeped in milk, and the fruit of choice in the late summer or early fall is black fig. We like to place these with the seeds up to showcase the beautiful pink flesh of the fruit as it comes out of the oven.

MAKES 8 SERVINGS

½ cup [100 g] granulated sugar, plus 2 Tbsp for the pan and more to top the figs

1⅓ cups [320 ml] whole milk

3 bay leaves (fresh or dried)

4 large eggs

2 tsp vanilla extract

Kosher salt

1⅓ cups [175 g] all-purpose flour

3 Tbsp unsalted butter, melted

2 lb [910 g] fresh black figs, stemmed and halved lengthwise

Confectioners' sugar, for serving

1. Preheat the oven to 350°F [180°C]. Generously butter a deep 9 in [23 cm] ceramic or glass pie dish. Evenly sprinkle about 2 Tbsp of granulated sugar in the dish.

2. In a small saucepan over medium-high heat, add the milk and bay leaves. While constantly stirring, bring the milk to a simmer, then quickly remove from the heat and steep the bay leaves in the milk while you make the batter.

3. In a medium bowl, whisk together the eggs, granulated sugar, vanilla, and a pinch of salt. Sift in half of the flour. Remove the bay leaves from the milk and alternate adding the milk and the remaining flour, whisking until combined. Add the melted butter and whisk until incorporated.

4. Arrange the sliced figs cut-side up around the bottom of the baking dish. Pour the batter over the figs, leaving about ⅛ in [4 mm] of rim at the top of the baking dish and the figs just peeking out through the batter and slightly floating. Sprinkle a pinch of sugar on top of the figs.

5. Bake for 1 hour until the top is lightly golden and firm. Let the dish come to room temperature and lightly dust with confectioners' sugar before serving. The clafoutis can be stored in an airtight container in the refrigerator for up to 3 days.

WINTER

Season of Slow Living

Our farm, the creamery, and West Marin slow way down in the winter, and I try my best to slow down with it. A neighboring rancher, Al Poncia of Stemple Creek Ranch, once said, "It isn't that time flies, it's just that it takes us a long time to fully live our lives." To optimize seasonal wellness, it makes sense that winter is about slowing down and living our lives fully.

In winter, the breeding season is over and lambing and kidding have yet to begin, so we spend time mending fences, deep cleaning, seeding pastures, reconditioning our roads, repairing our barns, and planning for spring. This is when I try to catch up on the many books I have yet to read and the embroidery projects that have been patiently waiting. We cook much more often, build fires, and stay at home. We eat and sell our cheeses that have been aging for eight months rather than the fresh cheeses that we make and sell in the spring.

A beautiful aspect of farm life is the focus on making things with your hands. In the last decade or so, it has been exciting to watch younger generations embrace the DIY ethos and return to making ceramics, cutting boards, clothes, jams, blankets, hats, refinished furniture, and so much more. The joy of gifting something you created is a tremendous feeling. We see the slow living of winter as the time to try out a new skill. In the culture most of us live in, we start to slow down in retirement. But we are making the case for slowing it all down seasonally, every year, so you can fully live your life.

TO START
TO START
TO START
TO START
TO START

TO START
TO START
TO START
TO START
TO START

Smoked Trout on Toast with Walnut-Lemon Gremolata

This open-faced sandwich is filling and flavorful, and any leftover gremolata can be used to garnish many other dishes. Lovely and inexpensive, this dish has a depth of flavor from the walnuts and can be made with tuna, chicken, or even simply goat cheese for a satisfying snack. We recommend making this dish with tinned trout packed in oil because it tends to be milder in flavor and softer in texture, but you can also use vacuum-sealed smoked trout and get a similar effect.

MAKES 2 SERVINGS

TOAST

2 slices sourdough bread, toasted

2 oz [55 g] fresh goat cheese (like our Liwa or Out Like a Lamb) or soft chèvre

4 oz [115 g] smoked trout or sardines

Freshly ground black pepper

GREMOLATA

½ cup [60 g] walnuts

8 to 10 fresh parsley sprigs, plus chopped parsley for garnish

3 garlic cloves

1 Tbsp extra-virgin olive oil

Zest and juice of 2 Meyer lemons (about 4 tsp zest and 3 Tbsp juice)

1 tsp kosher salt

½ tsp freshly ground black pepper

1. To make the toast, spread the toasted bread with a thin layer of cheese. Set aside.

2. To make the gremolata by hand, in a dry medium pan over medium-low heat, toast the walnuts, about 5 minutes, and set aside to cool slightly. Scatter the walnuts on a large wooden cutting board along with the parsley and garlic and begin chopping them together until the garlic is minced, the walnut pieces are finely chopped, and the parsley is finely chopped. Add this mixture to a small bowl and gently stir in the olive oil, lemon juice and zest, salt, and pepper. Alternatively, you can add the walnuts, parsley, garlic, salt, and pepper to a small food processor and blitz together, adding the oil and lemon juice and zest as you go, tasting occasionally to adjust the seasoning.

3. To serve, top the cheese toasts with thin slices of trout and dot generously with the gremolata. Add a flourish of chopped parsley and pepper. Serve immediately. We recommend eating this dish the same day it's made.

Roasted Beet Salad with Teleeka and Pancetta

In 1997, Vasco opened in Mill Valley. It was a small, casual, and affordable Italian restaurant on the corner of two main streets. In a town that was quickly transforming from a hippie enclave to a much fancier place, Vasco still offered a small-town vibe. When I was growing up, I loved going there, and I always ordered the same thing: the roasted beet salad. It was made with arugula, red beets, and pancetta, with a little fresh pesto spread on the plate. Sadly, Vasco closed its doors in 2020, so this is my homage to my favorite dish from our neighborhood joint.

MAKES 4 SERVINGS

6 beets, a mix of golden and red, peeled

4 oz [115 g] pancetta (we like Molinari), thinly sliced

3 Tbsp pesto

2 Tbsp balsamic vinegar

2 heads Little Gem lettuce, leaves torn into bite-size pieces

4 oz [115 g] triple cream cheese (like our Teleeka), sliced or crumbled

Crusty sourdough bread, for serving

1. Preheat the oven to 425°F [220°C].

2. Wrap the beets individually in aluminum foil and bake for 1 hour. Let cool. This can be done in advance, and the beets can be unwrapped and refrigerated for up to 2 days. Once cool, slice the beets into ½ in [13 mm] wedges.

3. Line a plate with paper towels. Place the pancetta in a dry sauté pan and turn on the heat to medium-high. Fry the pancetta until golden brown, about 5 minutes. Remove the pancetta with a slotted spoon and place on the paper towel–lined plate to drain, leaving the remaining fat in the pan. Add the beets to the pan along with the pesto and vinegar. Lower the heat to low and toss gently to coat the beets as they warm up, about 3 to 5 minutes.

4. To serve, in a large salad bowl, toss the beets with the lettuce. Divide among plates and top each salad with cheese. Serve immediately with crusty sourdough bread. We recommend eating this dish the same day it's made.

Little Gem and Watermelon Radish Salad with Koto'la

This crisp winter salad is peppery and sharp from endive, sweet from the pistachios, and beautifully fresh from the fennel and radish. It's simple to throw together and a wonderful departure if you're in a salad rut. The bright contrast of pink, green, and white colors in your bowl will surely boost your mood in the midst of winter.

1. To make the dressing, in a large salad bowl, whisk together the olive oil, vinegar, mustard, and orange zest and juice.
2. To make the salad, toss the croutons into the dressing to coat. Add the lettuce, endive, fennel, and radishes to the bowl and toss to coat. Sprinkle with the salt and pepper and toss again. Top with the pistachios, cheese, and roughly chopped fennel fronds. Serve immediately. We recommend eating this dish the same day it's made.

MAKES 4 SERVINGS

DRESSING

½ cup [120 ml] extra-virgin olive oil

1 Tbsp apple cider vinegar

1 tsp Dijon mustard

1 orange, zested and juiced (about 2 tsp zest and 3 Tbsp juice)

SALAD

2 cups [120 g] sourdough croutons

2 heads Little Gem lettuce, leaves torn into bite-size pieces

3 small red and green endive, cut into ½ in [13 mm] pieces

1 fennel bulb, sliced into half-moons, fronds roughly chopped and reserved

3 watermelon radishes, peeled and sliced into thin triangles

½ tsp kosher salt

¼ tsp freshly ground black pepper

¼ cup [30 g] shelled pistachios, lightly toasted

1 oz feta-style cheese (like our Koto'la)

FARM-STAY *Fun*

EMMA AUGUST (AGE 12), FARMSTAY GUEST

Toluma Farms is the most magical place for people of all ages. Every time we return, we are greeted with new and truly amazing experiences. These experiences range from a cheesemaking class to bathing baby goats and so much more. There are so many things we love to do outside to get everybody off the couch and enjoying nature. We love the beautiful walk around the property ending at what we like to call "Goat Hill," where we play with the cutest and cuddliest goats you could ever hope to meet.

And who could forget the trampoline?! My three siblings and I could spend hours on that trampoline doing nothing but bouncing and pleading for our dad to "skyrocket" us. The hot tub is the perfect place to warm up on a chilly evening. The chickens provide endless fresh eggs; the garden has so many delicious herbs, edible flowers, and juicy strawberries; and the goats make delectable cheeses to enjoy during our stay.

The staff is always so kind, never annoyed with our constant flow of questions and somehow always ready with an answer. And, of course, my personal favorite: going to the milking parlor, where we learn so many new things, things I would have never even thought about googling until we stumbled upon this magical place. Goats are the best!!! On top of all the special and amazing things you can do at Toluma Farms, the house is gorgeous and has the most amazing library and wall of board games I have ever seen. Thank you so much for all of our truly unforgettable stays at the farm!

(Emma and her three siblings, parents, and grandparents have traveled from Dallas three years in a row to stay at the farm.)

MAINS
MAINS
MAINS
MAINS
MAINS

MAINS
MAINS
MAINS
MAINS
MAINS

Wild Mushroom and Teleeka Quiche

In the middle of the rainy season, when mushrooms are abundant, incorporating them in a quiche makes for an elegant and warming dish that celebrates their diverse flavors and textures. The mix of wild mushrooms here could include fresh lion's mane, mini bella, trumpet, or cremini mushrooms, and morels either fresh or dried and soaked to reconstitute. This dish is appropriate for any meal of the day. It is rich and creamy, and the homemade pie crust is earthy and wonderful. We prefer to make the pie crust by hand, but you can use a food processor if you prefer.

MAKES 6 SERVINGS

CRUST

1½ cups [210 g] all-purpose flour, plus more for dusting

1 tsp kosher salt

¼ cup [55 g] cold unsalted butter, cut into 1 in [2.5 cm] pieces

¼ cup [55 g] shortening, chilled

¼ cup [60 ml] ice water

FILLING

2 Tbsp extra-virgin olive oil

1 leek, white and light green parts, chopped

4 cups [400 g] mixed wild mushrooms, finely chopped

1 Tbsp kosher salt, plus more for seasoning

6 large eggs, at room temperature

1 cup [240 ml] heavy cream

¼ cup [60 ml] whole milk

½ tsp freshly ground black pepper

4 oz [115 g] triple cream cheese (like our Teleeka)

1 Tbsp roughly chopped fresh thyme leaves

1. To make the crust, in a large mixing bowl, sift together the flour and salt. Add the pieces of butter and quickly toss it throughout the flour mixture to coat. Using your fingers, squeeze and work the butter into the flour until the mixture becomes somewhat shaggy in appearance. Add the shortening and continue to mix with your fingers until the dough forms a coarse mixture with some larger pieces remaining. Slowly add the ice water and mix with your hands until the dough just comes together into a ball.

2. Transfer the dough to a lightly floured work surface and form it into a disk. Wrap the disk in plastic wrap or beeswax wrap and refrigerate for 1 hour. Meanwhile, preheat the oven to 350°F [180°C].

3. Roll out the chilled dough on the floured work surface to form a 12 to 13 in [30.5 to 33 cm] circle. Lay the dough gently into a 9 in [23 cm] pie pan and top with parchment paper and dry beans or pie weights. Blind bake the crust for 20 to 25 minutes until slightly golden. Remove the weights and let the crust cool while you make the filling.

4. To make the filling, in a large sauté pan over medium heat, warm the olive oil. Add the leek and sauté for 8 to 10 minutes until softened and caramelized. Add the mushrooms and a pinch of salt and sauté for about 5 minutes until the mushrooms have reduced in size and are soft, browned, and fragrant.

5. In a large mixing bowl, crack the eggs and whisk to combine the yolks and whites. Add the heavy cream, milk, salt, and pepper. Pour the mushroom mixture into the bowl and stir. Pour the custard mixture into the pie crust, dot with the cheese, and sprinkle with half of the thyme. Bake for 1 hour until the center is set and no longer wobbly.

6. To serve, sprinkle the quiche with the remaining thyme. We recommend eating this dish the same day it's made, but it can be stored in an airtight container in the refrigerator for up to 3 days.

Whipped Liwa BLT

There seems to be no end to the variations you can do with a BLT. Some people are religious about a BLT requiring white bread, mayonnaise, and iceberg lettuce; others insist on layers of spice and assertive flavors to modernize this diner classic into an avant-garde garnish for a Bloody Mary. We're somewhere in the middle of the road on this one. We believe BLTs should always be on lightly toasted and thinly sliced sourdough bread. We say, why use iceberg lettuce if you can get your hands on Little Gem leaves? And because we have plenty of fresh goat cheese daily at Toluma Farms, we must swap out the mayonnaise for a light spread of Liwa, olive oil, and fresh lemon zest.

MAKES 1 SANDWICH

2 oz [55 g] fresh goat cheese (like our Liwa)

1 tsp extra-virgin olive oil

1 tsp grated lemon zest

¼ tsp kosher salt

2 thin slices sourdough bread, lightly toasted

3 slices crisp cooked bacon

3 to 4 Little Gem lettuce leaves

1 beefsteak tomato, thinly sliced and seasoned with salt and pepper

3 to 4 fresh basil leaves

1. In a small bowl using a small whisk or fork, vigorously mix the cheese, olive oil, lemon zest, and salt until it has a lightly whipped texture. Spread on one side of each bread slice. On top of one of the slices, layer on the bacon, lettuce, and seasoned tomatoes. Top with the basil and remaining slice of bread. To serve, cut the sandwich in half diagonally and enjoy immediately.

Roasted Dungeness Crab with Citrus Herb Butter

This dish is a Christmas Eve tradition for my family. It makes for a meal you will never forget—one that may soon become a tradition in your family as well. We are fortunate that we have to travel only a few miles north of the farm to Bodega Bay to get delicious crab. Although the copious amounts of melted spicy citrus butter and the garlicky herb bath drenching the crab will likely ruin your fancy holiday napkins, it's completely worth it. All pretense goes away when the steaming platter is brought to the table. Dive in with your hands and have extra napkins at the ready, as well as a light green salad and fresh sourdough bread. You will never want to eat crab any other way.

Be sure to buy fully cooked, cleaned, and cracked crabs. If they're frozen, you will need to defrost them a day before preparing them.

MAKES 4 SERVINGS

½ cup [110 g] plus 2 Tbsp unsalted butter

½ cup [120 ml] extra-virgin olive oil

6 Tbsp [90 g] minced garlic

2 Tbsp minced shallot

¼ cup [10 g] chopped fresh thyme

1 tsp red pepper flakes

4 large Dungeness crabs, cooked, cleaned, and cracked (about 4¼ lb [2 kg])

2 oranges, zested and juiced (about ½ cup [120 ml] juice and 2 tsp zest)

1 lemon, zested and juiced (about 3 Tbsp juice and 1½ tsp zest)

½ cup [120 ml] white wine

¼ cup [10 g], chopped fresh parsley, plus 2 Tbsp for garnish

¼ cup [60 ml] lemon juice

1. Preheat the oven to 450°F [230°C].

2. In a large, heavy enameled or cast-iron Dutch oven over medium-high heat, melt the ½ cup [110 g] of butter with the olive oil until it browns slightly and is fragrant. Lower the heat to medium and stir in the garlic, shallot, 2 Tbsp of the thyme, and the red pepper flakes. Add the cracked crabs and stir to combine. Cover and roast the crabs until heated through, stirring once, about 15 minutes.

3. To serve, using tongs, transfer the crabs to a platter. To the pot, add the orange juice and zest, lemon juice and zest, wine, parsley, and the remaining 2 Tbsp of thyme. Bring the mixture to a rolling boil until it is reduced by half, 5 to 7 minutes. Remove from the heat, add the remaining 2 Tbsp of butter, and stir. Pour the sauce over the crab and sprinkle with remaining 2 Tbsp of parsley to garnish.

    Any leftovers can be refrigerated in an airtight container for up to 3 days.

Fennel, Avocado, and Chicory Salad with Citrus and Atika

This is the perfect winter salad. There is a beautiful whimsy to a pink salad—no one expects it, especially in the dead of winter, but that's exactly when you'll find pink chicory at your farmers' market. The contrast of the green avocado, blood oranges, and rosy leaves is delightful, and the brown butter dressing adds a warm and earthy sweetness.

MAKES 4 SERVINGS

BROWN BUTTER DRESSING

2 Tbsp salted butter

1 Tbsp extra-virgin olive oil

1 Tbsp minced shallot

Juice of 1 blood orange

1 tsp kosher salt

½ tsp freshly ground black pepper

SALAD

1 avocado, sliced

Leaves from 2 large heads pink chicory (1 to 1¼ lb [450 to 570 g])

1 fennel bulb, thinly sliced on a mandoline

1 blood orange, peeled and sliced into disks

3 oz [85 g] Parmesan-style cheese (like our Atika), shaved

Freshly ground black pepper

1. To make the brown butter dressing, in a small saucepan over medium-low heat, melt the butter. Once it starts foaming, swirl the pan slowly and continue heating until the butter is slightly browned and aromatic, about 2 minutes. Remove from the heat and add the olive oil, shallot, blood orange juice, salt, and pepper and whisk to emulsify. Pour into a salad bowl.

2. To make the salad, add the avocado to the salad bowl and toss in the dressing to coat. Add the chicory leaves and fennel.

3. To serve, toss the salad and portion it onto plates. Top each with orange slices, cheese shavings, and pepper. Serve immediately. We recommend eating this dish the same day it's made.

Doug's Balsamic Fig and Tomato Chicken

This is the first dish that Jessica's mother-in-law taught Doug to make when he was ready to leave home for college. She knew he was a bit nervous in the kitchen, and this dish was simple enough for him to master but also elegant and impressive for a college student. This was also the first dish he cooked for Jessica when they moved in together. Nearly a decade later, they still make this dish and enjoy the incredible flavors and the nostalgia it brings.

MAKES 4 SERVINGS

- 2 Tbsp extra-virgin olive oil
- 4 bone-in, skin-on chicken thighs
- 1 tsp kosher salt
- ½ tsp freshly ground black pepper
- 1 cup [160 g] cherry tomatoes, stemmed
- 6 to 8 dried figs, halved
- 3 fresh thyme sprigs, plus fresh thyme leaves, for garnish
- 3 Tbsp balsamic vinegar

1. In a large cast-iron skillet over medium-high heat, warm the olive oil until shimmering but not smoking. Pat the chicken thighs dry and sprinkle with the salt and pepper. Add the chicken to the skillet, skin-side down, and brown, about 5 minutes. Turn the chicken and add the tomatoes, figs, and thyme. Drizzle with half of the balsamic vinegar and cover to finish cooking the chicken, about 10 minutes.

2. Check that the chicken is cooked through; it should register 170°F [76°C] on an instant thermometer. At this point, the tomatoes should have begun to burst and the dried figs should be very soft. Using tongs, transfer the chicken to a platter.

3. To the skillet over medium-low heat, add the remaining vinegar and stir, scraping the bottom of the skillet, further mashing the tomatoes, and reducing the vinegar, about 5 minutes. To serve, pour the tomato and fig sauce over the chicken and garnish with fresh thyme. This dish can be stored in an airtight container in the refrigerator for up to 3 days.

Kimchi, Kenne, and Bacon Pajeon

This recipe is contributed by chef Nicole Dobarro.

Chef Nicole Dobarro is Korean and was born and raised in Honolulu, Hawaii. Her innovative dishes combine her heritage with a twist (very San Francisco), like these pajeon (pancakes) that we can't stop eating. Her culinary journey has taken her to cooking at the contemporary Chinese restaurant Mister Jiu's and French restaurant Nico's. Nicole hosted a wonderful picnic on the farm for Seotdal Geumeum, which is the Lunar New Year in South Korea. Nicole and her chef husband, Marty Siggins, own and operate Hey Mauma out of Sausalito. They offer fresh, delivered meals inspired by Korean wisdom for nourishing postpartum healing.

MAKES 3 SERVINGS

½ cup [70 g] all-purpose flour

1 Tbsp potato starch

½ tsp baking soda

½ tsp kosher salt

¾ cup [180 ml] soda water

2 cups [300 g] kimchi, roughly chopped (we like Min-Hee Hill Gardens at the Marin Farmers' Market)

1 bunch green onions, cut into ¼ in [6 mm] pieces, plus chopped green onions for garnish

½ yellow onion, sliced ¼ in [6 mm] thick

5 strips cooked bacon, finely diced

6 oz [170 g] soft-ripened goat cheese (like our Kenne), chilled in the freezer and finely diced

¼ cup [60 ml] vegetable oil

1. In a large bowl, mix together the flour, potato starch, baking soda, and salt. Add the soda water and mix until there are no lumps. Add the kimchi, green onions, onion, bacon, and cheese and mix well.

2. In a 10 in [25 cm] nonstick pan over medium heat, add enough oil to coat the surface. Pour the batter into the pan to form a 3 to 4 in [7.5 to 10 cm] round. Cook until the bottom is crispy, about 6 to 8 minutes. Flip the pancake and cook the other side until crispy, another 6 to 8 minutes, adding more oil if needed. Transfer the pancake to a wire rack and repeat with the remaining batter.

3. To serve, stack the pancakes on a serving platter and garnish with green onions. Serve immediately. The pancakes can be stored in an airtight container in the refrigerator for up to 2 days.

Lamb and Pork Sausage Cassoulet

This is quintessential old-world peasant food, and we love to make this in the dead of winter. Cassoulet follows the same set of rules as most traditional stews: You brown the meat, soften the vegetables, add the beans and aromatics, then simmer in a low oven or covered over a low flame for an hour. Let this simplified version of the classic combination of pork, lamb, rosemary, and white beans warm you up on a cold winter night.

MAKES 6 SERVINGS

2 Tbsp extra-virgin olive oil

⅓ lb [150 g] pork belly or pancetta, diced

3 lamb sausages (9 oz [255 g]), casings removed

1 leek, white and light green parts, chopped (about 10 oz [280 g])

1 celery stalk, diced

15 to 20 garlic cloves, minced

8 oz [230 g] roughly chopped carrots

2 fresh rosemary sprigs

4 to 6 fresh thyme sprigs

2 bay leaves

60 oz [1.7 kg] cannellini or navy beans, soaked overnight and drained

4 to 6 small Roma tomatoes, quartered

4 cups [960 ml] chicken stock

2 Tbsp kosher salt

2 tsp freshly ground black pepper, plus more to serve

1 oz [30 g] Parmesan-style cheese (like our Atika), grated

1. In a large braising pot or Dutch oven over medium heat, warm the olive oil. Add the pork belly and brown, stirring occasionally, until most of the fat is rendered and the meat is slightly charred, about 3 to 4 minutes. Add the lamb sausage and brown until slightly crumbled and very fragrant, about 5 minutes. With a slotted spoon, transfer the pork and lamb to a plate.

2. To the pot, add the leek, celery, and half of the garlic. Lower the heat to medium-low and stir constantly until the leek is soft and slightly caramelized, about 5 minutes. Lower the heat to low, add the remaining garlic and the carrots, and sauté for 8 to 10 minutes until fragrant but not brown.

3. Gather the rosemary, thyme, and bay leaves and tie them together with kitchen twine into a bouquet garni. Return the meat to the pot and add the beans, tomatoes, chicken stock, and bouquet garni. Stir, then add the salt and pepper and cover the pot. Increase the heat to medium and cook until everything comes to a simmer. Simmer gently for approximately 1 hour or until the beans are cooked through. Remove the bouquet garni.

4. To serve, ladle the cassoulet into shallow dishes. Top with the cheese and fresh pepper to serve. This dish can be stored in an airtight container in the refrigerator for up to 5 days.

SWEET
SWEET
SWEET
SWEET
SWEET

SWEET
SWEET
SWEET
SWEET
SWEET

Campari Hazelnut Cake

This one-bowl cake is extremely versatile: You can easily make it with navel or blood oranges, lemons, or, our personal favorite, Meyer lemons. Blood oranges typically yield less zest and juice than lemons, so the suggested quantities for each are listed. We love a heavily perfumed, citrusy Campari cake, so we tend to err on the side of extra zest, but this is a personal preference; you can dial it up or down depending on what you like.

This is not a particularly sweet dessert, which means it's lovely any time of day, especially with tea or coffee. You can add additional sweetness by drizzling more of the radiant pink glaze over each slice.

1. Preheat the oven to 350°F [180°C]. Grease a 9 in [23 cm] round springform pan and line it with parchment paper cut to fit the bottom.

2. To make the cake, in a large mixing bowl, combine the citrus zest with the baking sugar and salt. Use your fingers to massage the zest into the sugar; the mixture will turn yellow/orange and should be the texture of wet sand.

3. Add the citrus juice to the sugar mixture along with the olive oil, eggs, and Campari. Whisk until smooth. Sift the hazelnut flour, all-purpose flour, baking powder, and baking soda right over the top of the sugar-egg mixture. Whisk well to combine.

4. Pour the mixture into the prepared pan and bake until the top is a light golden color and the cake begins to slightly pull away from the sides, about 45 minutes. Cool on a wire rack.

5. To make the citrus Campari glaze, in a small bowl, combine the confectioners' sugar with the citrus juice and the Campari. Whisk until smooth. Drizzle some of the glaze over the top of the cooled cake and pour the rest into a carafe.

6. To serve, sprinkle on the hazelnuts, offering the carafe of glaze on the side. The unglazed cake will keep for up to 3 days, wrapped tightly and stored in a cool, dry place. Store the glaze in an airtight container in the refrigerator for up to 7 days.

MAKES 8 SERVINGS

CAKE

5 Tbsp grated Meyer lemon or blood orange zest (from 3 to 5 lemons or 4 to 6 oranges)

1 cup [200 g] baking sugar or superfine sugar

½ tsp kosher salt

3 Tbsp Meyer lemon or blood orange juice

¾ cup [180 ml] extra-virgin olive oil

3 large eggs

2 Tbsp Campari

1½ cups [150 g] hazelnut flour

1 cup [140 g] all-purpose flour (substitute gluten-free flour if desired)

½ tsp baking powder

¼ tsp baking soda

CITRUS CAMPARI GLAZE

2 cups [240 g] confectioners' sugar

2 Tbsp Meyer lemon or blood orange juice, with pulp

1 Tbsp Campari

GARNISH

½ cup [50 g] toasted and finely chopped hazelnuts

Tomales Farmstead Creamery Soufflé

Whenever I make a soufflé I remember a quote from one of my favorite movies of all time, *Sabrina*: "A woman happily in love, she burns the soufflé. A woman unhappily in love, she forgets to turn on the oven." If you're able to remember to turn on the oven and to pay close attention to your timer, this cheesy soufflé should turn out beautifully. We recommend using room temperature butter in the molds, and if you don't have a traditional soufflé dish, you can make this in any dish that has tall, straight sides. It looks exceptionally beautiful coming out of the oven in a deep shiny saucepan.

MAKES 6 SERVINGS

6 Tbsp [85 g] unsalted butter, finely diced, plus more for greasing the pan

8 oz [230 g] grated Parmesan-style cheese (like our Atika), plus more for dusting the pan

6 Tbsp [60 g] all-purpose flour

½ tsp kosher salt

1½ cups [360 ml] half-and-half

6 large eggs, separated

2⅔ oz triple cream cheese (like our Teleeka), diced

⅛ tsp kosher salt

Pinch of freshly ground black pepper

1. Preheat the oven to 400°F [200°C].

2. Butter the inside of a 3 qt [2.8 L] soufflé dish. (The batter can also be divided evenly among 1 cup [240 ml] individual molds.) Using parchment paper, create a paper collar around the outside of the soufflé dish and tie with cotton kitchen twine to secure in place. This will help the soufflé rise evenly. Sprinkle the inside of the buttered dish with some Parmesan-style cheese and set aside.

3. In a 2 qt [1.9 L] saucepan, add the flour, salt, and half-and-half. Cook over medium heat, whisking constantly, until smooth and very thick. Transfer the hot mixture to a large mixing bowl and add the butter in small cubes, stirring with a wooden spoon to melt the butter. Set the béchamel mixture aside to cool slightly. The butter will gradually get absorbed.

4. Add the egg yolks to the béchamel one at a time, mixing well after each addition with the wooden spoon. Add the triple cream cheese, salt, and pepper and taste to check the seasoning.

5. In a large bowl or in a stand mixer fitted with a whisk attachment, add the egg whites. Start beating slowly and then increase the speed to medium-high and whip until stiff peaks form, about 4 minutes. Fold the egg whites into the béchamel mixture carefully so that they do not deflate but are completely absorbed.

6. Pour the mixture into the prepared soufflé dish and place on a baking sheet. Turn the oven broiler on and place the soufflé under the broiler for 3 to 4 minutes until a light crust forms. Turn the oven temperature back to 400°F [200°C] and bake for 20 to 25 minutes. Carefully remove from the oven and place the entire soufflé dish on the table to serve immediately. You may portion out the soufflé on individual plates using a large spoon. This dish is best enjoyed immediately after it comes out of the oven and should not be stored.

Giving Thanks: The Team at Toluma Farms and Tomales Farmstead Creamery

Someone recently asked me: Knowing what I know now, would I start this farm if I had it to do over again? The answer was an immediate yes. I am able to reply quickly because of the extraordinary people David and I have met along the way. We have been fortunate to connect with so many individuals who share our passion for every piece of the farm. They pour their hearts and souls into the care of the soil, goats, sheep, and cheese. The farm has had the good fortune to attract talented herd managers, land stewards, apprentices, volunteers, milkers, and cheesemakers. Besides their hard work, they have brought innovative ideas and deep compassion for small-scale regenerative farming.

Beyond our immediate team are other circles of community. There are the farmers who get up at 3 or 4 a.m. and drive their fruit, veggies, meats, and cheese to the farmers' markets. For almost a decade, my husband, who still has his day job as a physician at UCSF, worked the Sunday Marin Farmers' Market (the highlight of his week). He loved that his fellow market vendors were always ready to help when needed, whether by holding down a tent that was blowing over or taking over the booth so another vendor could take a quick break. That's the sort of camaraderie we need more of in today's world.

Then there are the customers—the diehard farmers' market shoppers who keep so many farms running. David is half Italian and half Jewish, so he's genetically programmed to feed others and watch them thoroughly enjoy that food. So there's no better place for him to be than handing out samples of our cheese on top of our neighbors' Acme bread with figs from Tomatero Farms on top.

Another community that we couldn't imagine living without are the farmers and food makers who surround us and have supported our journey for decades. Loren and Lisa of fourth-generation Stemple Creek, who welcomed us to town immediately and readily shared their experience of climate-friendly farming and running a food business. Emily and Guido, who operate True Grass Farms, a short walk across the street, and host a Thursday-night potluck for anyone who wants to swing by and are quick to lend a hand in a crisis. And the Furlong family, who have assisted us with every riparian and water project, offering skills that we ourselves do not possess.

Then there are the many cheese companies that didn't see us as a threat bent on taking a share of their business but instead subscribed to the idea that a rising tide raises all boats. Without the people behind Cowgirl Creamery, Redwood Hill Farm, Nicasio Valley Cheese Company, Point Reyes Farmstead Cheese Company, and Straus Creamery, we would not be in business today.

As you can see, we didn't do any of this alone. My biggest piece of advice to someone starting a farm or food business is to heavily collaborate with those around you. Go to the rancher meetings and sit on local boards to get to know a diverse range of people. Reach out for advice and seek out help for the skills you don't have. And give back to your community in exchange for all that you receive. The value of collaboration in our community was never more evident than during the height of COVID. Like others, we hunkered down at home, which was our farmhouse. Our adult children and their partners came in from Seattle and New York, and we had some wonderful months together. We supplied cheese and milk to our neighbors, and they supplied us with meat and veggies. We didn't need a store; we only needed each other.

Acknowledgments

I would like to express my gratitude to the village behind this book. Thank you to my husband, David, for his unwavering support and infectious enthusiasm for restoring anything that needs restoring; this time it happened to be a farm. This book would not exist without your unshakeable belief in me, your Capricorn partnership, your glass-half-full outlook, humor, and deep love—for that, I am endlessly grateful.

To my human kids, Josy, Emmy, Nico, and Megan: You are my greatest motivation. Each of you inspires me every day to do better. This book is for you, and in its small way I hope it inspires you to continue to explore your own imaginations and create your own epic stories. I love you more than words on a page can express.

My life has been blessed with brave and strong women. To my mother, Lucinda, my grandmothers, Millie and Delilah, and my great-grandmothers, Opal and Jewel: Thank you for showing me by example how to care for the earth, my communities, and, most important, myself. You are the backbone of all my ideas and endeavors.

Thank you to my core female friendship crew of over three decades for your unwavering support for every adventure I embark upon, even something as crazy as starting a goat and sheep dairy. Your encouragement during moments of doubt and your ability to make me laugh, even on the toughest days, have been invaluable. Thank you for being my sounding board and my nonjudgmental cheerleaders.

Entering the world of farming and sustainable food brought me more than just a deeper connection to the natural world; it brought me the most wonderful web of humans who are my closest friends and mentors. Your work not only nourishes our community but also fosters a deeper connection to the earth and to one another. Despite the challenges you face, you continue to bring your passion and values to your work, always prioritizing people and health over profit. Thank you for being stewards of the land, champions of making great food produced locally, and educating others about what you practice.

The immense joy and learning I have experienced on the farm over the last two decades is due to the incredible individuals who have chosen to join our team. I feel lucky to call many of them my friends and teachers. You've inspired me through your profound care for all living things, from bacteria and fungi in the soil to each goat, sheep, dog, cat, and chicken as well as the vibrant ecosystem of plants, birds, coyotes, and waterways. Your creativity, hard work, and care have provided me with hope when the world can feel hopeless. The planet is lucky to have you as its future caretakers. Please don't give up.

Lastly, thank you to my creative conspirator on this fun project, Jessica. You not only are an idea woman but you also see every glorious detail to the end. You provided the team with so much delicious nourishment on this journey. I am so grateful you found your way to the farm. I look forward to more collaborative creativeness in our future!

Tamara

This book is the manifestation of a long-held dream, and I have so many people to thank who have always supported my learning, creativity, exploration, and confidence to try new things. First and foremost, thank you to Tamara for welcoming me as a volunteer on the farm, supporting this vision from the very beginning, and agreeing to create this book together! Thank you for opening up the farmhouse to me to test new recipes, prepare meals for the staff and guests, and keep experimenting with the Tomales Farmstead Creamery cheeses. Thank you to the farm staff for taking care of all of the leftovers and taking good care of the goats, the sheep, and the land.

Looking back, I want to thank my mom and grandparents for always encouraging me to see what's happening in the kitchen, try new things, and create lasting traditions through food. I want to thank the amazing chefs, home cooks, and cookbook writers who have inspired me to try new techniques, use my intuition in the kitchen, and, most important, to cook with love.

I want to thank my incredible husband, Doug, who originally encouraged me to reach out to Toluma Farms to volunteer and has supported me in more ways than I can count—devouring every recipe I test and encouraging me to keep painting, to keep cooking, to keep trying, and to always remember to rest and enjoy our blessed and beautiful life here in Marin.

I want to thank my daughter, Mary, for always showing so much interest in what I'm making, and staying happy and patient in the baby carrier while I cook and bake. I hope this book inspires you to pursue your passions, creative interests, and know that the world is ready for your gifts.

Lastly I want to thank my friends and relatives who were so supportive—sending me recipe ideas, reviewing my work, and cheering me on—as well as the farmers and chefs who provided stories, guidance, and inspiration to celebrate the bounty of this region.

Jessica

Thank you to Chronicle Books for saying yes to *Feasts on the Farm*. We want to thank pioneering Nan McEvoy and our friend Nion McEvoy for keeping the magic of print and books alive.

Since this was our first book, we had a lot to learn, and fortunately we were in great hands from start to finish. Thank you, Tyrrell Mahoney, president of Chronicle Books, and Sarah Billingsley, publishing director for Food, Lifestyle, and Art, for listening to our pitch and encouraging us to move forward.

Thank you to the talented people who really made this book the book we envisioned. To our editor, Dena Rayess, and assistant editor, Alex Galou, for their patience as we learned while doing. Thanks to Chronicle Books's designer, Rachel Harrell, for listening to us and executing. We really lucked out with talented photographer and rancher Katie Newburn, prop stylist Natalia Poltoratzky of Cosita Creative, and food stylist and chef Christine Wolheim, along with her extremely hardworking and talented team of food stylists Allison Fellion, Huxley McCorkle, and Jessie Boom. You all made the magic happen!

We are fortunate to be making cheese in the San Francisco Bay Area, where chefs seek out delicious food that is made ethically and with equitable practices. We have no PR budget; thus we rely on the chefs and cheesemongers to find us, and they certainly do. These are the chefs who spend their weekends shopping at the farmers' markets and develop relationships with their food sources. Several of these chefs contributed recipes to this cookbook, and we thank them for prioritizing delicious food made locally.

What can you do to make a difference for your planet? Eating is an agricultural act. Know your farmers, and buy local to uplift your entire community. Shop at your local farmers' market and think about the climate impact on each thing you buy.

Land Acknowledgment

The Coast Miwok have lived in relation to this landscape since time immemorial. Prior to European colonization, more than six hundred village sites existed between the Bodega Bay region and southern Marin County, comprising one of the most densely populated areas in all of North America. Their heritage and culture are deeply intertwined with the land itself.

Today, the Coast Miwok people are working to restore healthy relationships with their traditional territories. We are grateful that we have partnered with the tribe to restore riparian, biodiverse areas on the farm, and we look forward to continued learning and partnerships as we tend to the wild.

Tamara and Jessica

Index

A

Afterglow Ice Cream, 24
Anderson, M. Kat, 19
Apple, Gravenstein, and Marmalade Grilled Cheese, 179
apricots
  Slow-Braised Lamb Shanks with Apricot Couscous, 73–75
  Stone Fruit and Flowers, 100
artichokes
  Goat Cheese Tatin, 154–58
  Savory Leek, Artichoke, and Atika Tart, 64
  Spring Vegetable Risotto with Kenne, Atika, and Liwa, 67–69
asparagus
  Asparagus and Teleeka Tart, 51
  Farmer's Panzanella Salad, 63
  Spring Vegetable Risotto with Kenne, Atika, and Liwa, 67–69
Atika (aged goat and sheep cheese), 35
  Asparagus and Teleeka Tart, 51
  Fennel, Avocado, and Chicory Salad with Citrus and Atika, 200
  Lamb and Pork Sausage Cassoulet, 205
  Lemon Bucatini with Atika, 119
  Liwa and Ham Frittata with Garden Lettuces, 66
  Meyer Lemon and Honey Pizza, 115
  Peach and Serrano Ham Pizza, 116
  Savory Leek, Artichoke, and Atika Tart, 64
  Spring Vegetable Risotto with Kenne, Atika, and Liwa, 67–69
  Teleeka and Summer Squash Lasagna, 120–23
  Tomales Farmstead Creamery Soufflé, 211
  Warm Olives with Atika, Herbs, and Lemon Zest, 153
August, Emma, 193
avocados
  Fennel, Avocado, and Chicory Salad with Citrus and Atika, 200
  Spicy Kale and Avocado Salad with Farro and Koto'la, 52
  Whipped Garden Herb and Liwa Mousse, 96

B

bacon. See also pancetta
  Kimchi, Kenne, and Bacon Pajeon, 204
  Maple-Crusted Butternut Squash with Liwa, 169
  Whipped Liwa BLT, 197
Balsamic Salad Pizza, 113
Banana Cachapa, Californios, 138
Banchero, Daniella, 120
Bay Laurel and Fig Clafoutis, 180
Bayou Sarah Farms, 24, 76
beans
  Lamb and Pork Sausage Cassoulet, 205
  Ross's Niçoise Salad, 173
  Smoky Baked Eggs with Chickpeas and Goat Cheese, 170
  Stemple Creek Rib-Eye with Rosemary Navy Beans, 131
beef
  Stemple Creek Rib-Eye with Rosemary Navy Beans, 131
Beet Salad, Roasted, with Teleeka and Pancetta, 190
Blackberry Corn Muffins, Mid-August, 137
BLT, Whipped Liwa, 197
Bossy (soft-ripened cow cheese), 33, 35
  Balsamic Salad Pizza, 113
Boyer, Terry Gamble, 139
Brandt, Julia, 55
bread. See also sandwiches
  croutons, 100
  Eggs Brouillés with Truffle Butter and Brioche, 57
  Farmer's Panzanella Salad, 63
  The Perfect Summer Cheese Board, 93
  Roasted Sungold Tomato and Teleeka Bites, 94
  Wild Mushroom, Thyme, and Liwa Toasts, 48
buffalo, 24, 76

C

Cachapa, Californios Banana, 138
cakes
  Campari Hazelnut Cake, 208
  Great-Grandma Opal's Chocolate Sheet Cake, 84
  Rhubarb Crumble Cake, 144
Caldera Farms, 139
Californios Banana Cachapa, 138
Campari Hazelnut Cake, 208
Campbell, Nick, 25, 103, 127–29
Carbon Cycle Institute, 19
carrots
  Lamb and Pork Sausage Cassoulet, 205
  Roasted Harissa Root Vegetables with Koto'la and Toasted Pumpkin Seeds, 160
  Slow-Braised Lamb Shanks with Apricot Couscous, 73–75
  Toluma Ploughman's Lunch, 108
Cashews, Kale Salad with Koto'la and, 97
Cassoulet, Lamb and Pork Sausage, 205
caviar
  Peas, Cheese, and Caviar!, 49
cheese. See also individual cheeses
  alternative view of, 127–29
  list of, 34–35
  making, 33–34
Cheese Board, The Perfect Summer, 93

The Cheesemaker's Proja, 103–5
chicken
  Doug's Balsamic Fig and Tomato Chicken, 203
  Roast Chicken with Fennel and Olives, 70
  Truffle Roast Chicken, 125
Chickpeas, Smoky Baked Eggs with Goat Cheese and, 170
chicory
  Fennel, Avocado, and Chicory Salad with Citrus and Atika, 200
Chocolate Sheet Cake, Great-Grandma Opal's, 84
Clafoutis, Bay Laurel and Fig, 180
climate-resilient farming, 19, 34
conservation, 18–19, 159
Cookies, Herbaceous Cocktail, 83
Cornish Game Hen Dijonnaise, 126
cornmeal
  The Cheesemaker's Proja, 103–5
  Mid-August Blackberry Corn Muffins, 137
Coughlin, Jenna, 24–25, 33
Couscous, Apricot, Slow-Braised Lamb Shanks with, 73–75
Crab, Roasted Dungeness, with Citrus Herb Butter, 199
Crackers, Smoked Trout on, with Liwa and Olives, 102
Croque Monsieur Canapé, Goat Cheese, 162
croutons, 100
cucumbers
  Farmer's Panzanella Salad, 63

D

dairies. See also individual dairies
  challenges faced by, 17
  history of, in Marin County, 16–17
  sustainability and, 20–21, 23
Dijon Grilled Cheese, 109
dinner parties, 89
Dobarro, Nicole, 204
Double 8 Dairy, 24, 115
Doug's Balsamic Fig and Tomato Chicken, 203

E

Eckhardt, Emily, 24
eggplant
  Goat Cheese Tatin, 154–58
eggs
  Eggs Brouillés with Truffle Butter and Brioche, 57
  Liwa and Ham Frittata with Garden Lettuces, 66
  Smoky Baked Eggs with Chickpeas and Goat Cheese, 170
  Toluma Ploughman's Lunch, 108

Tomales Farmstead Creamery Soufflé, 211
El Greco (brined cow cheese), 33, 35
The Cheesemaker's Proja, 103–5
endive
   Endive, Pear, and Kenne Salad with Toasted Walnuts and Honey, 54
   Little Gem and Watermelon Radish Salad with Koto'la, 192

F

Faber, Phyllis, 18
farmers' markets, 25, 89
Farmer's Panzanella Salad, 63
farming. *See also* dairies; *individual farms*
   allure of, 40, 45
   climate-resilient, 19, 34
   regenerative, 23–25
   sustainable, 18–21, 23–25
   women and, 148–49
Farro, Spicy Kale and Avocado Salad with Koto'la and, 52
fennel
   Cornish Game Hen Dijonnaise, 126
   Farmer's Panzanella Salad, 63
   Fennel, Avocado, and Chicory Salad with Citrus and Atika, 200
   Little Gem and Watermelon Radish Salad with Koto'la, 192
   Roast Chicken with Fennel and Olives, 70
   Roasted Harissa Root Vegetables with Koto'la and Toasted Pumpkin Seeds, 160
   Slow-Braised Lamb Shanks with Apricot Couscous, 73–75
   Spring Vegetable Risotto with Kenne, Atika, and Liwa, 67–69
Fibershed Learning Center, 31
figs
   Bay Laurel and Fig Clafoutis, 180
   Doug's Balsamic Fig and Tomato Chicken, 203
fish
   Ross's Niçoise Salad, 173
   Smoked Trout on Crackers with Liwa and Olives, 102
   Smoked Trout on Toast with Walnut-Lemon Gremolata, 189
flowers, edible
   Heirloom Tomato and Nasturtium Galette, 110
   Liwa Basque Cheesecake with Honeyed Nectarines and Nasturtiums, 143
   Stone Fruit and Flowers, 100
Frittata, Liwa and Ham, with Garden Lettuces, 66
fruit. *See also individual fruits*
   The Perfect Summer Cheese Board, 93
   Stone Fruit and Flowers, 100

G

Galette, Heirloom Tomato and Nasturtium, 110
goats
   dairy foods from, 28–29
   grazing habits of, 20–21, 28, 30, 165
   sociability of, 28, 30
   vegetation management and, 28, 30, 139–40
Gore, Al, 28
Great-Grandma Opal's Chocolate Sheet Cake, 84
Gruyère cheese
   Goat Cheese Croque Monsieur Canapé, 162

H

ham. *See also* prosciutto
   Goat Cheese Croque Monsieur Canapé, 162
   Liwa and Ham Frittata with Garden Lettuces, 66
   Peach and Serrano Ham Pizza, 116
Harvest Lentil Soup with Fried Liwa and Butter, 174
Hazelnut Cake, Campari, 208
herbs
   Herbaceous Cocktail Cookies, 83
   Warm Olives with Atika, Herbs, and Lemon Zest, 153
   Whipped Garden Herb and Liwa Mousse, 96

I

Ice Cream, Tahitian Vanilla–Cardamom Sheep Milk, 134
indigenous people, ecological knowledge of, 19

J

Jablons, Emmy Hicks, 58

K

kale
   Kale Salad with Cashews and Koto'la, 97
   Spicy Kale and Avocado Salad with Farro and Koto'la, 52
Kaufmann, Obi, 19
kefir
   The Cheesemaker's Proja, 103–5
   Crisp Popovers with Goat Kefir and Mulberry Rhubarb Compote, 80
Kenne (soft-ripened goat cheese), 34
   Endive, Pear, and Kenne Salad with Toasted Walnuts and Honey, 54
   Kimchi, Kenne, and Bacon Pajeon, 204
   Spring Vegetable Risotto with Kenne, Atika, and Liwa, 67–69
Kimchi, Kenne, and Bacon Pajeon, 204
Koto'la (brined goat cheese), 35
   Charred Watermelon Salad with Basil and Koto'la, 99
   The Cheesemaker's Proja, 103–5
   Farmer's Panzanella Salad, 63
   Kale Salad with Cashews and Koto'la, 97
   Lemon Bucatini with Atika, 119
   Little Gem and Watermelon Radish Salad with Koto'la, 192
   Roasted Harissa Root Vegetables with Koto'la and Toasted Pumpkin Seeds, 160
   Slow-Braised Lamb Shanks with Apricot Couscous, 73–75
   Smoky Baked Eggs with Chickpeas and Goat Cheese, 170
   Spicy Kale and Avocado Salad with Farro and Koto'la, 52
   Toluma Ploughman's Lunch, 108
Kuhn, Moira, 149

L

La Folie, 154
lamb
   Lamb and Pork Sausage Cassoulet, 205
   Marinated Lamb Chops, 72
   Slow-Braised Lamb Shanks with Apricot Couscous, 73–75
Lasagna, Teleeka and Summer Squash, 120–23
leeks
   Potato Leek Soup with Smoked Pepper and Liwa Crema, 164
   Savory Leek, Artichoke, and Atika Tart, 64
Left Bank Brasserie, 154
lemons
   Campari Hazelnut Cake, 208
   Lemon Bucatini with Atika, 119
   Meyer Lemon and Honey Pizza, 115
   Meyer Lemon and Rosemary Panna Cotta, 81
   Roasted Dungeness Crab with Citrus Herb Butter, 199
   Smoked Trout on Toast with Walnut-Lemon Gremolata, 189
Lentil Soup, Harvest, with Fried Liwa and Butter, 174
lettuce
   Balsamic Salad Pizza, 113
   Little Gem and Watermelon Radish Salad with Koto'la, 192
   Liwa and Ham Frittata with Garden Lettuces, 66
   Roasted Beet Salad with Teleeka and Pancetta, 190
   Ross's Niçoise Salad, 173
   Whipped Liwa BLT, 197
Little Wing farm stand, 67
Liwa (fresh goat cheese), 34
   Asparagus and Teleeka Tart, 51
   Californios Banana Cachapa, 138
   Eggs Brouillés with Truffle Butter and Brioche, 57
   Goat Cheese Croque Monsieur Canapé, 162
   Goat Cheese Tatin, 154–58
   Harvest Lentil Soup with Fried Liwa and Butter, 174
   Heirloom Tomato and Nasturtium Galette, 110

Liwa and Ham Frittata with Garden Lettuces, 66
Liwa Basque Cheesecake with Honeyed Nectarines and Nasturtiums, 143
Maple-Crusted Butternut Squash with Liwa, 169
Peas, Cheese, and Caviar!, 49
Potato Leek Soup with Smoked Pepper and Liwa Crema, 164
Savory Leek, Artichoke, and Atika Tart, 64
Smoked Trout on Crackers with Liwa and Olives, 102
Smoked Trout on Toast with Walnut-Lemon Gremolata, 189
Spring Vegetable Risotto with Kenne, Atika, and Liwa, 67–69
Toasted Liwa Muffin with Dill and Honey, 163
Whipped Garden Herb and Liwa Mousse, 96
Whipped Liwa BLT, 197
Wild Mushroom, Thyme, and Liwa Toasts, 48
Lucero, Priscilla and Curtis, 149

M

MacKenzie, Jenny, 33
Maine Farmland Trust, 24
Maple-Crusted Butternut Squash with Liwa, 169
Marin Agricultural Land Trust (MALT), 18, 159
Marin Resource and Conservation District, 18
Marin Roots Farms, 149
McEvoy, Nan, 89
McEvoy Ranch, 89
McKibben, Bill, 28
Mid-August Blackberry Corn Muffins, 137
Mister Jiu's, 204
Mousse, Whipped Garden Herb and Liwa, 96
muffins
    Mid-August Blackberry Corn Muffins, 137
    Toasted Liwa Muffin with Dill and Honey, 163
Muir, John, 19
Mulberry Rhubarb Compote, Crisp Popovers with Goat Kefir and, 80
mushrooms
    Goat Cheese Tatin, 154–58
    Spring Vegetable Risotto with Kenne, Atika, and Liwa, 67–69
    Wild Mushroom, Thyme, and Liwa Toasts, 48
    Wild Mushroom and Teleeka Quiche, 196

N

Natural Resource and Conservation Services (NRCS), 18
nectarines
    Liwa Basque Cheesecake with Honeyed Nectarines and Nasturtiums, 143
    Stone Fruit and Flowers, 100
Niçoise Salad, Ross's, 173
Nico's, 204

O

Oliver, Mary, 41
olives
    Goat Cheese Tatin, 154–58
    Roast Chicken with Fennel and Olives, 70
    Ross's Niçoise Salad, 173
    Smoked Trout on Crackers with Liwa and Olives, 102
    Warm Olives with Atika, Herbs, and Lemon Zest, 153
oranges
    Campari Hazelnut Cake, 208
    Fennel, Avocado, and Chicory Salad with Citrus and Atika, 200
    Little Gem and Watermelon Radish Salad with Koto'la, 192
    Roasted Dungeness Crab with Citrus Herb Butter, 199

P

Pajeon, Kimchi, Kenne, and Bacon, 204
Palace Market, 67
pancetta
    Farmer's Panzanella Salad, 63
    Lamb and Pork Sausage Cassoulet, 205
    Roasted Beet Salad with Teleeka and Pancetta, 190
Panna Cotta, Meyer Lemon and Rosemary, 81
Panzanella Salad, Farmer's, 63
parsnips
    Roasted Harissa Root Vegetables with Koto'la and Toasted Pumpkin Seeds, 160
Passot, Roland, 154
pasta
    Lemon Bucatini with Atika, 119
    Teleeka and Summer Squash Lasagna, 120–23
Peach and Serrano Ham Pizza, 116
pears
    Endive, Pear, and Kenne Salad with Toasted Walnuts and Honey, 54
peas
    Farmer's Panzanella Salad, 63
    Peas, Cheese, and Caviar!, 49
    Spring Vegetable Risotto with Kenne, Atika, and Liwa, 67–69
Piccino, 120
pine nuts
    Slow-Braised Lamb Shanks with Apricot Couscous, 73–75
pistachios
    Little Gem and Watermelon Radish Salad with Koto'la, 192
pizzas
    Balsamic Salad Pizza, 113
    Meyer Lemon and Honey Pizza, 115
    Peach and Serrano Ham Pizza, 116
    Sourdough Pizza Crust, 114
Place and Purpose podcast, 19
Point Blue Conservation, 18
pomegranates
    Kale Salad with Cashews and Koto'la, 97
Poncia, Al, 185
Poncia, Loren and Lisa, 131
Popovers, Crisp, with Goat Kefir and Mulberry Rhubarb Compote, 80
pork. *See also* bacon; ham; pancetta; prosciutto
    Lamb and Pork Sausage Cassoulet, 205
potatoes
    Potato Leek Soup with Smoked Pepper and Liwa Crema, 164
    Ross's Niçoise Salad, 173
    Toluma Ploughman's Lunch, 108
    Truffle Roast Chicken, 125
Proja, The Cheesemaker's, 103–5
prosciutto
    Dijon Grilled Cheese, 109
    Stone Fruit and Flowers, 100
puff pastry
    Asparagus and Teleeka Tart, 51
    Savory Leek, Artichoke, and Atika Tart, 64
pumpkin seeds
    Maple-Crusted Butternut Squash with Liwa, 169
    Roasted Harissa Root Vegetables with Koto'la and Toasted Pumpkin Seeds, 160

Q

Quiche, Wild Mushroom and Teleeka, 196

R

radishes
    Little Gem and Watermelon Radish Salad with Koto'la, 192
    Roasted Harissa Root Vegetables with Koto'la and Toasted Pumpkin Seeds, 160
Raine, Audrey, 20
raspberries
    Crisp Popovers with Goat Kefir and Mulberry Rhubarb Compote, 80
Redwood Hill Creamery, 33
regenerative farming, 23–25
rhubarb
    Crisp Popovers with Goat Kefir and Mulberry Rhubarb Compote, 80
    Rhubarb Crumble Cake, 144
rice
    Spring Vegetable Risotto with Kenne, Atika, and Liwa, 67–69
ricotta cheese
    Homemade Ricotta, 123
    Teleeka and Summer Squash Lasagna, 120–23
Risotto, Spring Vegetable, with Kenne, Atika, and Liwa, 67–69

Roland, Sarah, 24, 76
romanesco
   Toluma Ploughman's Lunch, 108
Ross's Niçoise Salad, 173
Route One Bakery and Kitchen, 93

S

salads
   Balsamic Salad Pizza, 113
   Charred Watermelon Salad with Basil and Koto'la, 99
   Endive, Pear, and Kenne Salad with Toasted Walnuts and Honey, 54
   Farmer's Panzanella Salad, 63
   Fennel, Avocado, and Chicory Salad with Citrus and Atika, 200
   Kale Salad with Cashews and Koto'la, 97
   Little Gem and Watermelon Radish Salad with Koto'la, 192
   Roasted Beet Salad with Teleeka and Pancetta, 190
   Ross's Niçoise Salad, 173
   Spicy Kale and Avocado Salad with Farro and Koto'la, 52
   Stone Fruit and Flowers, 100
sandwiches
   Dijon Grilled Cheese, 109
   Goat Cheese Croque Monsieur Canapé, 162
   Gravenstein Apple and Marmalade Grilled Cheese, 179
   Smoked Trout on Toast with Walnut-Lemon Gremolata, 189
   Whipped Liwa BLT, 197
Sarris, Greg, 19
Sausage Cassoulet, Lamb and Pork, 205
sheep, 31, 134
Silva, Adriana, 148–49
Small Shed, 113
Soufflé, Tomales Farmstead Creamery, 211
soups
   Harvest Lentil Soup with Fried Liwa and Butter, 174
   Potato Leek Soup with Smoked Pepper and Liwa Crema, 164
Sourdough Pizza Crust, 114
spinach
   Harvest Lentil Soup with Fried Liwa and Butter, 174
   Liwa and Ham Frittata with Garden Lettuces, 66
   Spring Vegetable Risotto with Kenne, Atika, and Liwa, 67–69
squash
   Goat Cheese Tatin, 154–58
   Maple-Crusted Butternut Squash with Liwa, 169
   Teleeka and Summer Squash Lasagna, 120–23
Steele, Clara, 16
Stemple Creek Ranch, 131, 185
Stowaway, Seth, 49
Straus, Ellen, 18
Straus, Vivien, 34
Students and Teachers Restoring A Watershed (STRAW), 18, 26
sustainability, 18–21, 23–25

T

Tahitian Vanilla–Cardamom Sheep Milk Ice Cream, 134
tarts
   Asparagus and Teleeka Tart, 51
   Goat Cheese Tatin, 154–58
   Savory Leek, Artichoke, and Atika Tart, 64
Teleeka (soft-ripened goat, sheep, and cow cheese), 35
   Asparagus and Teleeka Tart, 51
   Dijon Grilled Cheese, 109
   Gravenstein Apple and Marmalade Grilled Cheese, 179
   Peach and Serrano Ham Pizza, 116
   Roasted Beet Salad with Teleeka and Pancetta, 190
   Roasted Sungold Tomato and Teleeka Bites, 94
   Stone Fruit and Flowers, 100
   Teleeka and Summer Squash Lasagna, 120–23
   Tomales Farmstead Creamery Soufflé, 211
   Wild Mushroom and Teleeka Quiche, 196
Three Shepherds Farm, 33
Toluma Farms
   apprenticeship program at, 23–25, 55, 76
   children and, 58, 193
   farmers' markets and, 89
   history of, 16
   restaurants and, 89
   as sanctuary for healing and renewal, 55
   seasons at, 26–27, 185
   sustainability practices at, 20–21
Toluma Ploughman's Lunch, 108
Tomales Farmstead Creamery, 33–35, 40, 93, 211
Tomatero Farms, 148–49
tomatoes
   Doug's Balsamic Fig and Tomato Chicken, 203
   Goat Cheese Tatin, 154–58
   Heirloom Tomato and Nasturtium Galette, 110
   Lamb and Pork Sausage Cassoulet, 205
   Roasted Sungold Tomato and Teleeka Bites, 94
   Ross's Niçoise Salad, 173
   Smoky Baked Eggs with Chickpeas and Goat Cheese, 170
   Sungold Tomato Sauce, 123
   Whipped Liwa BLT, 197
Trevino, Tina, 160, 165
trout
   Smoked Trout on Crackers with Liwa and Olives, 102
   Smoked Trout on Toast with Walnut-Lemon Gremolata, 189
True Grass Farms, 66, 89
truffle butter
   Eggs Brouillés with Truffle Butter and Brioche, 57
   Truffle Roast Chicken, 125
tuna
   Ross's Niçoise Salad, 173
turnips
   Slow-Braised Lamb Shanks with Apricot Couscous, 73–75

V

Van de Bovenkamp, Nico, 119
Vandendriessche, Anne Marie, 33
vanilla
   Tahitian Vanilla–Cardamom Sheep Milk Ice Cream, 134
Vasco, 190
vegetables. *See also individual vegetables*
   Roasted Harissa Root Vegetables with Koto'la and Toasted Pumpkin Seeds, 160
   Spring Vegetable Risotto with Kenne, Atika, and Liwa, 67–69
   Toluma Ploughman's Lunch, 108
Verdone, Lily, 159

W

walnuts
   Endive, Pear, and Kenne Salad with Toasted Walnuts and Honey, 54
   Smoked Trout on Toast with Walnut-Lemon Gremolata, 189
watermelon
   Charred Watermelon Salad with Basil and Koto'la, 99
   Little Gem and Watermelon Radish Salad with Koto'la, 192
wildfires, 30, 139–40
Wolf, Kate, 26
women-run farms, 148–49
Wool Shed, 31

Y

yogurt
   Crisp Popovers with Goat Kefir and Mulberry Rhubarb Compote, 80
   Marinated Lamb Chops, 72

Z

Zlot, Andrew, 24
zucchini
   Goat Cheese Tatin, 154–58
   Teleeka and Summer Squash Lasagna, 120–23

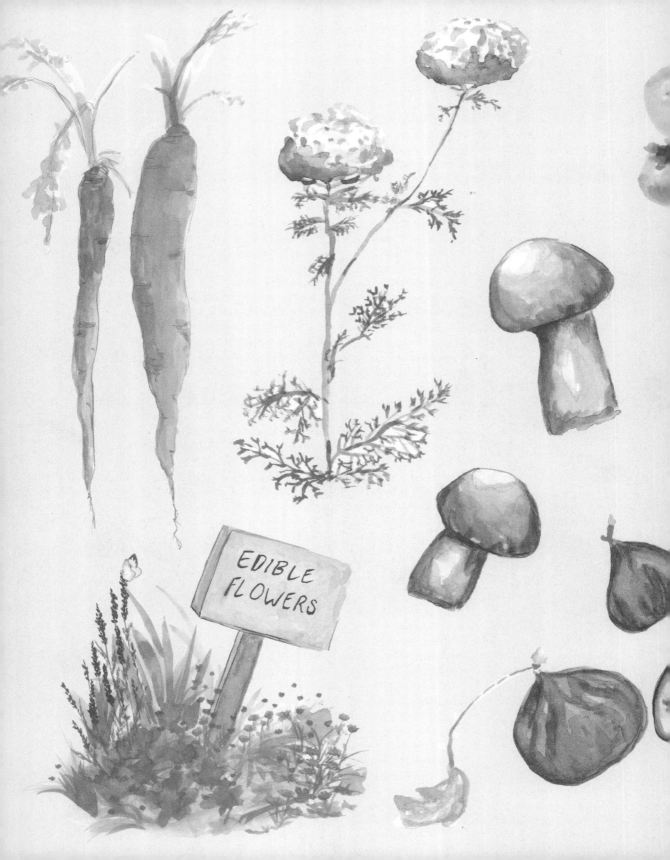